Walking to Olympus:
An EVA Chronology

David S. F. Portree and
Robert C. Treviño

NASA History Office
Office of Policy and Plans
NASA Headquarters
Washington DC 20546

Monographs in Aerospace History Series #7
October 1997

Foreword

One of the most significant activities conducted in space takes place when human beings depart their spacecraft and travel about and perform work in a spacesuit. Extravehicular activities (EVA) require some of the most complex technical skills, sophisticated technologies, and human capabilities of all missions undertaken in space. The first of these EVAs took place on 18 March 1965 during the Soviet Union's Voskhod 2 orbital mission when cosmonaut Alexei Leonov first departed the space-craft in Earth orbit to test the concept. In June of 1965, during the flight of Gemini 4, Edward White II, performed the first EVA by an American. Since that time hundreds of hours have been amassed by humans conducting EVAs in both Earth orbit and on the lunar surface. Between that time and April 1997, when Jerry Linenger conducted an EVA with Vladimir Tsibliyev as part of International Space Station Phase I, 154 EVAs have been undertaken.

These total EVAs have not only accomplished significant work in space, work impossible through any other means, but also yielded enormous knowledge, skills, and experience among the astronaut and cosmonaut corps about how to perform meaningful work beyond the confines of Earth's atmosphere. *Walking to Olympus: An EVA Chronology*, by David S.F. Portree and Robert C. Treviño, is a comprehensive chronicle of all of the EVAs conducted since the dawn of the space age. Because history is so important in helping to chart the direction for the future, this monograph's publication is especially significant because the building of the International Space Station near the end of this century will require many more hours of EVA than has been previously undertaken altogether. One of our goals for publishing this monograph at this time is to help inform officials and the general public of what EVAs have been done before, what they accomplished, and what hurdles had to be surmounted to accomplish them.

This is the seventh publication in a new series of special studies prepared by the NASA History Office. The **Monographs in Aerospace History** series is designed to provide a wide variety of investigations relative to the history of aeronautics and space. These publications are intended to be tightly focused in terms of subject, relatively short in length, and reproduced in an inexpensive format to allow timely and broad dissemination to those interested in aerospace history. Suggestions for additional publications in the **Monographs in Aerospace History** series are welcome.

ROGER D. LAUNIUS
Chief Historian
National Aeronautics and Space Administration
September 11, 1997

Contents

Acronyms and Abbreviations

A7L	Apollo suit, 7th model, International Latex Corporation	ELSS	Extravehicular Life Support System
A7LB	Apollo suit, 7th model, International Latex Corporation, B variant	EMU	Extravehicular Mobility Unit
		ERA	French deployable space structure
ACCESS	Assembly Concept for Construction of Erectable Space Structures	ESA	European Space Agency
		ESEF	European Space Exposure Facility
ALSA	Astronaut Life Support Assembly	Eureca	European Retrievable Carrier
ALSEP	Advanced Lunar Scientific Experiment Package	EVA	Extravehicular Activity
		F	Fahrenheit
AMU	Astronaut Maneuvering Unit	FAS	Fixed Airlock Shroud
ASEM	Assembly of Station by EVA Methods	fps	feet per second
		FSM	Functional Service Module
ASIPE	Axial Scientific Instrument Protective Enclosure	ft	foot, feet
		g	gravity, gravities
ATDA	Augmented Target Docking Adapter	G4C	Gemini suit, 4th model, by David Clark Company
ATM	Apollo Telescope Mount		
BET	Beam Erection Tether	GHSP	Goddard High Speed Photometer
BLSS	Buddy Life Support System	GRO	Gamma Ray Observatory
BOSS	Soviet visible light communications system	GSFC	Goddard Space Flight Center
		HHMU	Handheld Maneuvering Unit
btu	British thermal unit	hr	hour(s)
C	centigrade	HST	Hubble Space Telescope
CDR	commander	HST SM-01	HST Servicing Mission-01
CETA	Crew and Equipment Translation Aid	HST SM-02	HST Servicing Mission-02
		HUT	Hard Upper Torso
CLIP	Crew Loads Instrumented Pallet	ISC	instrument-science compartment
cm	centimeter(s)	in	inch(es)
CM	Command Module	ISS	International Space Station
CMP	Command Module pilot	IV	Intravehicular
COSTAR	Corrective Optics Space Telescope Axial Replacement	JSC	Johnson Space Center
		kg	kilogram(s)
CPD	Crew Propulsive Device	km	kilometer(s)
CSR	Crew Self Rescue	KRT-10	Soviet space-based radio telescope
deg	degree(s)	kW	kilowatt(s)
dia	diameter	lb	pound(s)
DCM	Display and Control Module	LCG	Liquid Cooling Garment
DM	Docking Module	LCVG	Liquid Cooling and Ventilation Garment
DTO	Development Test Objective		
EASE	Experimental Assembly of Structures through EVA	LEVA	Lunar Extravehicular Visor Assembly
EASEP	Early Apollo Scientific Experiment Package	LM	Lunar Module
		LMP	Lunar Module pilot
ECC	Electronic Cuff Checklist	LRV	Lunar Roving Vehicle
ECU	Electronic Control Unit	m	meter(s)
EDFT	EVA Development Flight Test	M	Modified

MEEP	Mir Environmental Effects Payload		tive Enclosure
MESA	Modular Equipment Stowage Assembly	RSU	Rate Sensing Unit
		SSA	Space Suit Assembly
MET	Modularized Equipment Transporter	SADE	Solar Array Drive Electronics
MFR	Manipulator Foot Restraint	SAFER	Simplied Aid for EVA Rescue
mi	mile(s)	SALC	Special Airlock Compartment
min	minute(s)	sec	second(s)
MIRAS	Mir Infrared Atmospheric Spectrometer	SEVA	stand-up extravehicular activity
		SIM	Scientific Instrument Module
mm	millimeter(s)	SIR-B	Shuttle Imaging Radar-B
MMS	Multimission Modular Spacecraft	SM	Service Module
MMU	Manned Maneuvering Unit	SNAP	Space Nuclear Applications Program
mo	month(s)		
MOMS	Mir Optoelectrical Multispectral Scanner	SPK	*Sredstvo Peredvizheniy Kosmonavtov* (cosmonaut maneuvering equipment)
MPESS	Mission Preculiar Equipment Support Structure		
		SPT	Science Pilot
mps	meters per second	SSF	Space Station Freedom
MSC	Manned Spacecraft Center	STS	Space Transportation System
MSFC	Marshall Space Flight Center	T	Transport
MSRE	Mir Sample Return Experiment	TGIF	Thank God It's Friday
MSS	Magnetic Sensing System	TM	Transport Modified
mW	megawatt(s)	TPAD	Trunnion Pin Attachment Device
NASA	National Aeronautics and Space Administration	TREK	space particle collector
		TsUP	*Tsentr Upravleniya Polyotami* (Soviet/Russian flight control center)
NPO	National Production Organization		
ODU	Salyut main propulsion system		
OKB-1	Special Design Bureau-1	TV	television
ORU	Orbital Replaceable Unit	URI	*Universalny Rabochy Instrument* or *Universalny Ruchnoj Instrument* (Universal Hand Tool)
PE	Principal Expedition		
PFR	Portable Foot Restraint		
PGA	Pressure Garment Assembly		
PIE	Particle Impact Experiment	URS	Soviet deployable assembly device
PLSS	Portable Life Support System	USAF	United States Air Force
PLSS	Primary Life Support System	UV	ultraviolet
PSA	Provisional Stowage Assembly	VDU	thruster package on Soviet Mir station
psi	pounds per square inch		
PWP	Portable Work Platform	VE	Visiting Expedition
rev	revolution(s)	WETF	Weightless Environmental Training Facility
RKK	Russian Space Corporation		
RMS	Remote Manipulator System	WFPC	Wide Field/Planetary Camera
rpm	revolutions per minute	wk	week(s)
RSA	Russian Space Agency	yr	year(s)
RSIPE	Radial Scientific Instrument Protec-		

Introduction

Spacewalkers enjoy a view of Earth once reserved for Apollo, Zeus, and other denizens of Mt. Olympus. During humanity's first extravehicular activity (EVA), Alexei Leonov floated above Gibraltar, the rock ancient seafarers saw as the gateway to the great unknown Atlantic. The symbolism was clear - Leonov stepped past a new Gibraltar when he stepped into space.

More than 32 years and 154 EVAs later, Jerry Linenger conducted an EVA with Vladimir Tsibliyev as part of International Space Station Phase I. They floated together above Gibraltar. Today the symbolism has new meaning - humanity is starting to think of stepping out of Earth orbit, space travel's new Gibraltar, and perhaps obtaining a new olympian view - a close-up look at Olympus Mons on Mars.

Walking to Olympus: An EVA Chronology chronicles the 154 EVAs conducted from March 1965 to April 1997. It is intended to make clear the crucial role played by EVA in the history of spaceflight, as well as to chronicle the large body of EVA "lessons learned."

Russia and the U.S. define EVA differently. Russian cosmonauts are said to perform EVA any time they are in vacuum in a space suit. A U.S. astronaut must have at least his head outside his spacecraft before he is said to perform an EVA. The difference is based in differing spacecraft design philosophies. Russian and Soviet spacecraft have always had a specialized airlock through which the EVA cosmonaut egressed, leaving the main habitable volume of the spacecraft pressurized. The U.S. Gemini and Apollo vehicles, on the other hand, depressurized their entire habitable volume for egress.

In this document, we apply the Russian definition to Russian EVAs, and the U.S. definition to U.S. EVAs. Thus, for example, Gemini 4 Command Pilot James McDivitt does not share the honor of being first American spacewalker with Ed White, even though he was suited and in vacuum when White stepped out into space.

Non-EVA spaceflights are listed in the chronology to provide context and to display the large number of flights in which EVA played a role. This approach also makes apparent significant EVA gaps - for example, the U.S. gap between 1985 and 1991 following the Challenger accident.

This NASA History Monograph is an edited extract from an extensive *EVA Chronology and Reference Book* being produced by the EVA Project Office, NASA Johnson Space Center, Houston, Texas. The larger work will be published as part of the NASA Formal Series in 1998.

The authors gratefully acknowledge the assistance rendered by Max Ary, Ashot Bakunts, Gert-Jan Bartelds, Frank Cepollina, Andrew Chaikin, Phillip Clark, Richard Fullerton, Steven Glenn, Linda Godwin, Jennifer Green, Greg Harris, Clifford Hess, Jeffrey Hoffman, David Homan, Steven Hopkins, Nicholas Johnson, Eric Jones, Neville Kidger, Joseph Kosmo, Alexei Lebedev, Mark Lee, James LeBlanc, Dmitri Leshchenskii, Jerry Linenger, Igor Lissov, James McBarron, Clay McCullough, Joseph McMann, Story Musgrave, Dennis Newkirk, James Oberg, Joel Powell, Lee Saegesser, Andy Salmon, Glen Swanson, Joseph Tatarewicz, Kathy Thornton, Chris Vandenberg, Charles Vick, Bert Vis, David Woods, Mike Wright, John Young, and Keith Zimmerman. Special thanks to Laurie Buchanan, John Charles, Janet Kovacevich, Joseph Loftus, Sue McDonald, Martha Munies, Colleen Rapp, and Jerry Ross. Any errors remain the responsibility of the authors.

The Chronology

1965

Voskhod 2 launch

March 18
1965 EVA 1
World EVA 1
Russian EVA 1
Duration: 0:24
Spacecraft/mission: Voskhod 2
Crew: Pavel Belyayev, Alexei Leonov
Spacewalker: Alexei Leonov
Purpose: Perform EVA ahead of U.S.; demonstrate feasibility of EVA

The U.S. response to the launch of Yuri Gagarin, the first human being to reach space (Vostok 1, April 1961) was President John Kennedy's call for a manned lunar landing by the end of 1969. The Gemini program was conceived to let NASA perfect skills needed for Apollo lunar flights, including EVA. Soviet leaders felt compelled to respond. Soyuz was under development, but would not be ready before Gemini, so Soviet engineers modified Vostok capsules to meet or beat some of Gemini's goals. The modified Vostok was called Voskhod. Immediately after reaching orbit in Voskhod 2, Leonov and Belyayev attached the EVA backpack to Leonov's Berkut ("Golden Eagle") suit. Berkut was a modified Vostok Sokol-1 intravehicular (IV) suit. A white metal backpack provided 45 min of oxygen for breathing and cooling. Oxygen vented through a relief valve into space, carrying away heat, moisture, and exhaled carbon dioxide. Suit pressure could be set at either 40.6 kpascal (5.88 psi) or 27.4 kpascal (3.97 psi). Belyayev then deployed and pressurized the Volga inflatable airlock. The airlock was necessary because Vostok/Voskhod avionics were cooled by cabin air and would overheat if the capsule was depressurized for EVA. Volga was designed, built, and tested in just 9 months beginning in mid-1964. At launch Volga fitted over Voskhod 2's hatch, extending 74 cm (29.6 in) beyond the spacecraft hull. The airlock comprised a metal ring 1.2-m (3.96-ft) wide fitted over Voskhod 2's inward-opening hatch; a double-walled fabric airlock tube with a deployed length of 2.5 m (8.25 ft); and a metal upper ring 1.2 m (3.96 ft) wide around the inward-opening airlock hatch 65 cm (26 in) wide. Volga's deployed internal volume was 2.5 cu/m (88.3 cu/ft). The fabric airlock tube was made rigid by about 40 airbooms clustered in three independent groups. Two groups were sufficient for deployment. The airbooms needed 7 min for full inflation. Four spherical tanks held sufficient oxygen to inflate the airbooms and pressurize the airlock. Two lights lit the airlock interior, and three 16-mm cameras - two inside the airlock and one outside on a boom mounted to the upper ring - were positioned to record the historic first spacewalk. Belyayev controlled the airlock from inside Voskhod 2, but a set of backup controls for Leonov was suspended on bungee cords inside the airlock. Leonov entered Volga, then Belyayev sealed Voskhod 2 behind him and depressurized the airlock. Leonov opened Volga's outer hatch and pushed out to the end of his 15.35-m (50.7-ft) umbilical. He later stated that the umbilical gave him tight control over his movements - an observation belied by subsequent U.S. EVA experience. Leonov reported looking down and seeing from the Straits of Gibraltar to the Caspian Sea. After Leonov returned to his couch, Belyayev fired pyrotechnic bolts to discard Volga. Sergei Korolev, Chief Designer at OKB-1 Design Bureau (now RKK Energia), stated after the EVA that Leonov could have remained outside for much longer than he did, while Mstislav Keldysh, "chief theoretician" of the Soviet space program and President of the Soviet Academy of Sciences, said that the EVA showed that future cosmonauts would find work in space easy. The government news agency TASS reported that, "outside the ship and after returning, Leonov feels well." However, post-Cold War Russian documents have revealed a different story. They report that Leonov's Berkut suit ballooned, making bending difficult. Because of this, Leonov was unable to reach the shutter switch on his thigh for his chest-mounted camera. He could not take pictures of Voskhod 2, nor was he able to recover the camera mounted on Volga which recorded his EVA for posterity. After 12 min Leonov reentered Volga. Recent accounts say that he violated procedure by entering the airlock head first, then got stuck sideways when he turned to close the outer hatch. This forced him to flirt with dysbarism (the "bends") by lowering his suit pressure so he could bend enough to free himself. Leonov recently revealed that he had a suicide pill he could have swallowed if he had been unable to ingress Voskhod 2 and Belyayev had been forced to leave him in orbit. Doctors reported that Leonov nearly suffered heatstroke - his core body temperature climbed 1.8 deg C (3.1 deg F) in 20 min - and Leonov stated that he was "up to his knees" in sweat, so that his suit sloshed when he moved. In an interview published in the *Soviet Military Review* in 1980, Leonov downplayed his difficulties and stated that "building manned orbital stations and exploring the Universe are inseparably linked with man's activity in open space. There is no end of work in this field."

"The First Egress of Man Into Space" (NASA TT F-9727), by Alexei Leonov, translation of "Pervyy vykhod chelovka v kosmicheskoye prostranstvo," presented at the XIVth International Astronautics Congress, Athens, September 13-18, 1965; "Orbital Castling: Life of Mir Station Will Be Prolonged Three More Years," *Segodnya* in Russian, May 11, 1995, p. 9. Translated in *JPRS Report, Central Eurasia: Space*, August 2, 1995 (FBIS-UST-95-030), p. 20; "Man In Open Space," *Soviet Military Review* interview, March, 1980, pp. 14-17. In *USSR Report: Space*, No. 11 (JPRS 78264), June 10, 1981, p. 19; "The Friendly Solar Wind," *Komsomol'skaya Pravda* in Russian, Alexei Leonov, March 18, 1983, p. 4. Translated in USSR Report: Space, No. 24 (JPRS 84161), August 22, 1983, pp. 11-13; "Voskhod 2 Flight Recalled," *Spaceflight*, June 1990, p. 193; *Astronautics and Aeronautics, 1965*, NASA SP 4006, 1966, pp. 132, 138.

March 19	Voskhod 2 landing
March 23	Gemini 3
June 3	Gemini 4 launch

June 3
1965 EVA 2
World EVA 2
U.S. EVA 1
Duration: 0:36
Spacecraft/mission: Gemini 4
Crew: James McDivitt, Edward White
Spacewalker: Edward White
Purpose: Demonstrate EVA feasibility; test HHMU

"It's the saddest moment of my life," said Ed White after his commander, James McDivitt, ordered him to return to the Gemini 4 spacecraft and conclude the first U.S. EVA. White's EVA was originally scheduled to begin about 3 hr after launch, during Gemini 4's second revolution about the Earth. However, an attempt to rendezvous with the expended upper stage of Gemini 4's Titan launch vehicle proved arduous (and ultimately unsuccessful) and EVA preparation required more time than expected, so the EVA was postponed to the third revolution. After wrestling with a jammed latch and pushing back the stiff spacecraft hatch, White stood in his seat and installed a 16-mm camera outside Gemini 4 to record the EVA, becoming slightly out of breath. He then used a Hand-Held Maneuvering Unit (HHMU) to leave the spacecraft. The device permitted only 20 sec of maneuvering before exhausting its compressed oxygen supply, but the test was judged a success. White inspected Gemini 4, then evaluated his umbilical, which turned out to be useful for limiting distance, but not for more precise maneuvering. In addition to supplying oxygen, the 7-m

Gemini 4, 1965 - Edward White II performs the first U.S. EVA. He wears a G4C space suit. Note the HHMU (top right) and White's VCM chestpack. (S65-30433)

(25-ft) umbilical carried communication and bioinstrumentation lines and a load-alleviating tether. The umbilical was covered by a thin layer of gold to protect it from the Sun's heat. An overglove that escaped from White's open hatch decayed from orbit within a few months and burned up in the atmosphere. White accidentally smeared McDivitt's window before returning to his seat and recovering the camera. He intentionally discarded his thermal overgloves and helmet sun visor before returning to Gemini 4. White remained comfortable in his Gemini G4C suit - he stated later that he was more comfortable during the EVA than at any other time during the flight - until the hatch refused to close. In the 5-min struggle to shut it he exceeded the cooling capacity of his chest-mounted Ventilation Control Module (about 1000 btu/hr), slightly fogging his visor. Sweat streamed into his eyes until the cabin was repressurized and he removed his helmet. Because the hatch proved difficult to close, it was not reopened as planned to discard EVA equipment. Joseph Shea, Director of the Apollo Project Office, National Aeronautics and Space Administration (NASA) Manned Spacecraft Center (MSC), stated later that White's EVA showed that Apollo Lunar Excursion Module crew members would be able to cross to the Apollo Command Module (CM) after ascent from the lunar surface if the two spacecraft could not dock. He also stated that engineers were considering wrapping the CM in cloth to protect it from particles deposited by the solid-fuel rockets which would jettison the Apollo launch escape tower. After reaching space, an astronaut would perform an EVA to peel off the cloth layer.

Summary of Gemini Extravehicular Activity, NASA SP 149, 1967; "Life Support Systems for Extravehicular Activity," Harold McMann, Elton Tucker, Marshall Horton, and Frederick Burns, in *Gemini Summary Conference*, NASA SP 136, February 1967, p. 73; *On the Shoulders of Titans: A History of Project Gemini*, Barton Hacker and James Grimwood, NASA SP 4203, 1977, pp. 246-250; *Project Gemini: A Chronology*, NASA SP 4002, 1969, pp. 200-202; "Gemini 4 Paves Way for Bolder Program," *Aviation Week & Space Technology*, June 7, 1965, p. 18; "Gemini 4 Success to Intensify Launch Pace," *Aviation Week & Space Technology*, June 14, 1965, p 83; "Gemini 4 Data to Aid Future Manned Flight," George Alexander, *Aviation Week & Space Technology*, June 21, 1965, p. 79; "Gemini 4 Medical Results Indicate Feasibility of 14-Day Spaceflights," Erwin Bulban, *Aviation Week & Space Technology*, June 21, 1965, p. 81; *Personal Logs*, Joseph McMann.

June 7	**Gemini 4 splashdown**
August 21-29	**Gemini 5**
December 4-18	**Gemini 7**
December 15-16	**Gemini 6**

1966

March 16	**Gemini 8**
June 3	**Gemini 9 launch**

June 5
1966 EVA 1
World EVA 3
U.S. EVA 2
Duration: 2:09
Spacecraft/mission: Gemini 9
Crew: Thomas Stafford, Eugene Cernan
Spacewalker: Eugene Cernan
Purpose: Conduct first complex EVA; test the AMU

The ease with which White adapted to EVA on Gemini 4 and the cancellation of David Scott's Gemini 8 EVA due to spacecraft problems encouraged EVA planners to schedule a full slate of activities for Eugene Cernan's Gemini 9 spacewalk. According to the mission plan, Gemini 9 would dock with the Augmented Target Docking Adapter (ATDA). Mission Pilot Cernan would emerge and retrieve an experiment package from the ATDA, then move to Gemini 9's aft adapter section, where the Astronaut Maneuvering Unit (AMU) was stowed. He would unfold its arms, attach himself to its integral life support system, unfasten his 7.5-m (25-ft) tether, and fly up to 45 m (150 ft) from Gemini 9. The crew reached the ATDA in orbit on June 3 to find that its launch shroud was still attached, preventing docking. Rendezvous efforts were tiring, so Command Pilot Thomas Stafford postponed the EVA for 24 hr. On this date Cernan emerged just before orbital dawn. He used an Environmental Life Support System (ELSS) chestpack similar to the one Scott would have used on Gemini 8, but lacked an HHMU. His G4C suit legs were modified with steel fabric and aluminized film layers to ward off heat from the AMU's hydrogen peroxide rockets. The thick, complex tether proved difficult to manage (Cernan called it "the snake"). Then Cernan moved to the AMU. Handrails, velcro pads, and loop foot restraints failed to help him control his movements. Cernan stated later that he "had to work continually against the pressure suit. . . I was devoting 50 percent of my workload just to maintain position." As he struggled, he broke off an experiment antenna mounted on Gemini 9 and tore the outer layers of his suit. His exertions exceeded the capacity of the ELSS to remove moisture, fogging his faceplate and blinding him. Cernan donned the AMU by touch, but Stafford called him back inside. Cernan also experienced "hot spots" on his back caused by sunlight striking torn places on his suit. After returning to Earth, Cernan conducted underwater neutral buoyancy simulations of his EVA in the Weightless Immersion Facility pool at NASA MSC. He reported that neutral buoyancy simulation nearly duplicated actual EVA conditions, helping to validate it as an EVA training tool.

Astronautics and Aeronautics, 1966, NASA SP 4007. pp. 202, 206-207, 218; *Summary of Gemini Extravehicular Activity*, NASA SP 149, 1967; "Life Support Systems for Extravehicular Activity," Harold McMann, *et al*, in *Gemini Summary Conference*, NASA SP 136, February 1967, p. 71-72; "Body Positioning and Restraints During Extravehicular Activity," David Schulz, *et al*, in *Gemini Summary Conference*, NASA SP 136, February 1967, pp. 79-90; *Project Gemini: A Chronology*, NASA SP 4002, 1969, pp. 245-246; *On the Shoulders of Titans: A History of Project Gemini*, NASA SP 4203, 1977, pp. 337-339.

June 6	Gemini 9 splashdown
July 18	Gemini 10 launch

July 19
1966 EVA 2
World EVA 4
U.S. EVA 3
Duration: 0:49
Spacecraft/mission: Gemini 10
Crew: John Young, Michael Collins
Spacewalker: Michael Collins
Purpose: SEVA to conduct photography

Gemini 10 docked with Agena 10 on July 18. Command Pilot John Young and Mission Pilot Michael Collins had to compensate for a trajectory error, leaving Gemini 10 with only half the fuel planned when the docking occurred. They then repeatedly fired Agena 10's engine to match orbits with Agena 8. Collins performed this Stand-up EVA (SEVA), the first of two EVAs planned for the mission, during the climb to Agena 8. He used a 70-mm camera to snap 22 images of the southern Milky Way in the ultraviolet part of the spectrum. As the docked Gemini 10 and Agena 10 spacecraft passed into daylight Collins took photos of a colored plate to determine if film accurately captured colors in space. Collins and Young then experienced eye irritation and smelled a strange odor in their suits. This was apparently caused by lithium hydroxide leaking into their helmets when both suit fans operated simultaneously. Lithium hydroxide in the Gemini life support system absorbed exhaled carbon dioxide.

Astronautics and Aeronautics, 1966, NASA SP 4007. pp. 243-244; *Summary of Gemini Extravehicular Activity*, NASA SP 149, 1967; *On the Shoulders of Titans: A History of Project Gemini*, NASA SP 4203, 1977, pp. 347-348.

July 20
1966 EVA 3
World EVA 5
U.S. EVA 4
Duration: 0:39
Spacecraft/mission: Gemini 10
Crew: John Young, Michael Collins
Spacewalker: Michael Collins
Purpose: Demonstrate translation between two spacecraft across open space; retrieve equipment from target spacecraft

As they matched orbits with Agena 8, Young and Collins separated Gemini 10 from Agena 10. Mindful of their scarce fuel, they rendezvoused with Agena 8 on this date about 90 min before the planned second EVA. They checked out their suits, prepared the cabin for depressurization, and wiped the inside of their helmet visors with an antifogging chemical. Collins then donned his ELSS chestpack and unstowed the umbilical which carried oxygen, communications, and nitrogen. The nitrogen, propellant for the HHMU, was stored under pressure in a tank in Gemini 10's adapter section. Freed from stowage, the umbilical filled the cockpit. Ground controllers told the astronauts to leave one of the two suit fans turned off to prevent lithium hydroxide from invading the suit loop again. Collins' second EVA was scheduled to last 90 min. He left the cabin, unfolded a handrail, and removed a micrometeoroid package attached to Gemini 10. He had to avoid the 16 thrusters Young used to keep the spacecraft close to Agena 8. The Gemini 10 hand controllers were modified to make piloting the spacecraft easier during EVA proximity operations while in a pressurized suit with restricted wrist mobility. This modification was later applied to the Apollo lunar rover. Collins then prepared to move to Agena 8 to retrieve a micrometeoroid package. He plugged a line hanging loose from his umbilical into the nitrogen connector on Gemini 10's adapter section. Collins described this task in his autobiographical book *Carrying the Fire*:

> my legs are flailing back and forth in response to the slightest torque that my arms put on the rail or the connector. . . I have missed on my first attempt to stab the connector with the fitting on the end of my umbilical. The sleeve on the fitting has sprung forward and must be recocked, but that is a two-handed operation. I let go of the rail for an instant, recock the sleeve and grab the rail again. In the process I swing wildly and bang up against the side of the spacecraft. John feels the commotion, and so does the Gemini's attitude control system, which reacts to this unwanted motion by firing thrusters.

Collins guided Young as he maneuvered Gemini 10 to within 3 m of the Agena, then Collins jumped to it. He gripped the Agena docking cone, dislodging a sharp-edged electric discharge ring which he feared might tear his suit or cut his umbilical. Young meanwhile kept track of a "3-body problem" involving Gemini 10, Collins, and Agena 8, all the while trying to keep sunlight from falling on Collin's ejection seat. The seat might have fired - taking Gemini 10's hatches with it - if it had been heated by the Sun for too long. Reaching the micrometeoroid package, Collins attempted to stop his forward motion, but his lower body momentum left him "turning lazy cartwheels somewhere above and to the left of everything that matters." He used the HHMU to stop his rotation, landed among the thrusters behind the Gemini 10 cockpit, and caught himself on his open hatch door. He then used the HHMU to move to the Agena. He retrieved the micrometeoroid package, but set the Agena gyrating, making it more difficult for Young to keep Gemini 10 close. He elected not to install a replacement package as planned. Gemini 10's fuel supply dropped dangerously low, forcing Young to call Collins back inside and cut short the EVA. Collins discovered that he had lost his camera, then the long umbilical caused problems again, preventing Young from seeing the control panel so he could report fuel usage to Houston and causing Collins to accidentally shut off the radio. They herded the umbilical into a bag with empty food containers and disused equipment and opened the hatch just long enough to toss it out. Collins' EVA was more successful than Cernan's, though he reported later that the "basic problem" of EVA remained that "without some sort of handholds or restraining devices, a large percentage of the astronaut's time is. . . devoted to torquing his body around until it is in the proper position to do some useful work." Gemini 10 was the first U.S. space mission preceded by underwater EVA training. Speaking in 1996, Young stated that he was under orders from Donald Slayton, Chief of the Astronaut Office, to get Collins back into the spacecraft if he became incapacitated. However, according to Young, "there was no way if anything happened to somebody going outside a Gemini that you could get them back in." The seat was too narrow and the pressure suit too stiff to put an EVA astronaut into the cockpit without his cooperation.

Carrying the Fire, An Astronaut's Journeys, Michael Collins, 1974, pp. 218-243; *On the Shoulders of Titans: A History of Project Gemini*, NASA SP 4203, 1977, pp. 349-350; *Summary of Gemini Extravehicular Activity*, NASA SP 149, 1967; interview, David S. F. Portree with John Young, June 13, 1996.

July 21	Gemini 10 splashdown
September 12	Gemini 11 launch

September 13
1966 EVA 4
World EVA 6
U.S. EVA 5
Duration: 0:38
Spacecraft/mission: Gemini 11
Crew: Charles Conrad, Richard Gordon
Spacewalker: Richard Gordon
Purpose: Demonstrate ability to perform a complex EVA

Mission Pilot Richard Gordon and Command Pilot Charles Conrad docked their spacecraft with Agena 11 on September 12, setting the stage for Gordon's first EVA. During the 107-min spacewalk, Gordon would attach a 30-m (100-ft) tether stowed on Agena 11 to Gemini 11's nose for an artificial gravity experiment. Also, he would retrieve the S9 nuclear emulsion package from Gemini 11's adapter section, and test the "golden slipper" foot restraint, an HHMU, and a torqueless power tool. Poor planning ensured that Gordon's EVA started badly. As dictated in the mission plan, he commenced EVA preparations 4 hr before scheduled EVA start. These required only 50 min. Gordon tested oxygen flow from the suit life support system and became uncomfortably warm because the life support oxygen cooling system heat exchanger could not be used - it was designed for vacuum operation. Just before opening the hatch he worked up a sweat trying to attach a visor to his helmet. Later he said that "I was pretty tired and had a pretty high heart rate before I ever opened the hatch." He then attempted a leap to Agena 11, missing and swinging on his 30-ft umbilical to Gemini 11's adapter section. Using the umbilical, Conrad pulled him back to the hatch for another try. This time Gordon succeeded in grasping handrails on the Agena docking adapter added following Collins' difficulties. However, Gordon still needed both hands to secure the tether to Gemini 11. He straddled the spacecraft nose as he had in zero-g aircraft simulations, but in space the G4C suit's internal pressure forced his legs together, pushing him away from the nose. He secured the tether while holding onto the handrail with one hand. Gordon moved back to the cockpit area to rest and Conrad ordered him back inside. An hour later, the astronauts opened the hatch and jettisoned loose equipment. Gordon said later that "a little simple task that I had done many times in training to the tune of about 30 seconds lasted about 30 minutes." After the flight Gordon said, "Gene Cernan warned me about this. . . I knew it was going to be harder [than on the ground], but I had no idea of the magnitude." Neutral buoyancy simulation was not yet a mandatory EVA training tool, so Gordon spent little time underwater preparing for his EVA. Gordon's experience encouraged Apollo lunar surface EVA astronauts to practice more in their suits.

"Gemini XI Crew Face the Press, Give Details of Flawless Flight." *Space News Roundup*, NASA Manned Spacecraft Center, September 20, 1966, pp. 1-2; *On the Shoulders of Titans: A History of Project Gemini*, NASA SP 4203, 1977, pp. 356, 360-362; *Summary of Gemini Extravehicular Activity*, NASA SP 149, 1967; *Astronautics and Aeronautics, 1966*, NASA SP 4007. pp. 301; interview, David S. F. Portree with John Young, June 13, 1996.

September 14
1966 EVA 5
World EVA 7
U.S. EVA 6
Duration: 2:08
Spacecraft/mission: Gemini 11
Crew: Charles Conrad, Richard Gordon
Spacewalker: Richard Gordon
Purpose: SEVA to perform ultraviolet astronomical photography and Earth photography

Like Collins before him, Gordon had few problems during his SEVA. He opened the hatch just before orbital sunset, installed the S13 ultraviolet astronomical camera, and took pictures of Orion and Antares. A short tether held him in the cabin, permitting him to use both hands. During the daylight pass Gordon performed "general photography," which included snapping pictures of Houston and Florida. During their pass over the Atlantic they had no photographic targets, so both astronauts fell asleep - a testimony to the relaxed pace of this EVA. The spacecraft again moved into darkness, and Gordon snapped more pictures of astronomical targets. Experiment S13 closeout and hatch closure were uneventful.

Astronautics and Aeronautics, 1966, NASA SP 4007, pp. 290; *On the Shoulders of Titans: A History of Project Gemini*, NASA SP 4203, 1977, pp. 365-366; *Summary of Gemini Extravehicular Activity*, NASA SP 149, 1967.

September 15   **Gemini 11 splashdown**

November 11   **Gemini 12 launch**

November 12
1966 EVA 6
World EVA 8
U.S. EVA 7
Duration: 2:18
Spacecraft/mission: Gemini 12
Crew: James Lovell, Edwin Aldrin
Spacewalker: Edwin Aldrin

Purpose: SEVA to familiarize Mission Pilot with EVA environment; conduct ultraviolet astronomical photography and Earth photography

EVA was a primary objective of Project Gemini; EVA requirements helped dictate that Gemini be a two-person spacecraft, and early plans had EVAs on every flight save the first. Therefore, it was with desperation that NASA reached the final flight of the Gemini program without a single complex EVA it could call an unqualified success. Great care was taken in training, planning, and providing handholds. Twelve mobility aids were added to the mission following Gordon's Gemini 11 difficulties, eight of which had not flown before in space. One new feature of the Gemini 12 mission plan was a relaxed SEVA designed to let Mission Pilot Edwin Aldrin become accustomed to his suit and equipment prior to the more demanding full-emergence EVA. Aldrin emerged in orbital daylight with an enthusiastic "Man, look at that!" and installed the S13 ultraviolet astronomical camera. He evaluated standup EVA dynamics until after dark, then performed astronomical photography. Shortly after the Gemini 12-Agena 12 combination flashed orange in the brief orbital dawn, Aldrin installed a camera to record his activities, then prepared for the next EVA by installing a handbar and unfolding a handrail. He changed a diffraction grating on the S13 camera and removed the S12 micrometeoroid package from behind the cockpit. Command Pilot James Lovell assisted Aldrin in his tasks. Just before orbital sunset he retrieved the EVA camera. Aldrin resumed astronomical photography as darkness fell again. He witnessed a second orbital sunrise before closing the hatch on his successful first EVA.

Summary of Gemini Extravehicular Activity, NASA SP 149, 1967; *On the Shoulders of Titans: A History of Project Gemini*, NASA SP 4203, 1977, pp. 377-378; interview, David S. F. Portree with John Young, June 13, 1996.

November 13
1966 EVA 7
World EVA 9
U.S. EVA 8
Duration: 2:09
Spacecraft/mission: Gemini 12
Crew: James Lovell, Edwin Aldrin
Spacewalker: Edwin Aldrin
Purpose: Demonstrate ability to perform complex EVA

Before flight Aldrin conducted five neutral buoyancy training sessions (not a large number by modern standards) in preparation for this EVA, in addition to the usual zero-g aircraft training. Aldrin also became accustomed to the relative immobility of the pressurized G4C suit in Thermal Vacuum Chamber B at NASA MSC. "Thermal vac" testing subsequently became a critical part of EVA training. To start this EVA, Aldrin moved to the Target Docking Adapter on Agena 12, where he used waist tethers to hold position. Attaching a tether on the Agena to Gemini 12 proved surprisingly easy with both hands free. He then moved back to the adapter section, where he slipped his feet into "golden slipper" foot restraints. He used waist tethers to position himself at a work station for testing representative tasks - he cut cables and fluid lines, fastened rings and hooks, connected and disconnected electrical and fluid connectors, tightened bolts, and stripped velcro. Aldrin's physical condition was closely monitored so that he could be advised to rest before fatigue developed. He moved to a similar work station on the Agena docking adapter where he tested an Apollo torque wrench with and without tethers. He wiped Lovell's window and observed thruster firings on Gemini 12, then closed the hatch on the world's first successful complex EVA.

Summary of Gemini Extravehicular Activity, NASA SP 149, 1967; *On the Shoulders of Titans: A History of Project Gemini*, NASA SP 4203, 1977, pp. 378.

November 14
1966 EVA 8
World EVA 10
U.S. EVA 9
Duration: 1:11
Spacecraft/mission: Gemini 12
Crew: James Lovell, Edwin Aldrin
Spacewalker: Edwin Aldrin
Purpose: SEVA to discard refuse; conduct ultraviolet astronomical photography

Aldrin's last Gemini 12 EVA - and the final EVA of the Gemini program - was anticlimactic, but helped confirm that U.S. EVA planners were on a sure footing going into the Apollo program. Aldrin jettisoned disused equipment just before orbital sunset, then performed ultraviolet photography. He photographed sunrise then stowed his gear and closed out the EVA. The *Summary of Gemini Extravehicular Activity* states that the Gemini 12 EVAs showed that

> all the tasks attempted were feasible when body restraints were used to maintain position. The results also showed that EVA workload could be controlled within desired limits by application of proper procedures. . . Finally, perhaps the most significant result was that underwater simulation duplicated the actual extravehicular actions and reactions with a high degree of fidelity. It was concluded that any task which could be accomplished readily in underwater simulation would have a high probability of success during the actual EVA.

Summary of Gemini Extravehicular Activity, NASA SP 149, 1967, pp. 3-26.

November 15	Gemini 12 splashdown

1967

April 23-24	Soyuz 1

1968

October 11-22	Apollo 7
October 26-30	Soyuz 3
December 21-27	Apollo 8

1969

January 14	Soyuz 4 launch
January 15	Soyuz 5 launch

January 16
1969 EVA 1
World EVA 11
Russian EVA 2
Duration: 0:37
Soyuz 4 crew: Vladimir Shatalov, Yevgeni Khrunov, Alexei Yeliseyev (launch to EVA)
Soyuz 5 crew: Boris Volynov, Yevgeni Khrunov, Alexei Yeliseyev (EVA to landing)
Spacewalkers: Yevgeni Khrunov, Alexei Yeliseyev
Purpose: Demonstrate EVA transfer between two spacecraft

This docking mission had EVA objectives similar to those planned for Apollo 9. Soyuz 4 launched first, and was the active vehicle in the docking with Soyuz 5. The news agency TASS stated that: ". . . there was a mutual mechanical coupling of the ships. . . and their electrical circuits were connected. Thus, the world's first experimental cosmic station with four compartments for the crew was assembled and began functioning. . ." The mission rehearsed elements of the Soviet piloted lunar mission plan. Moscow TV carried the cosmonauts' EVA preparations live. Khrunov and Yeliseyev put on their Yastreb ("hawk") suits in the Soyuz 5 orbital module with aid from Commander Boris Volynov. Yastreb suit design commenced in 1965, shortly after Leonov's difficult EVA. Leonov served as consultant for the design process, which was complete during 1966. Suit fabrication and testing occurred in 1967, but the Soyuz 1 accident in April of that year and Soyuz docking difficulties (Soyuz 2-Soyuz 3, October 1968) delayed use in space until Soyuz 4-Soyuz 5. To prevent the suit ballooning which contributed to Leonov's EVA difficulties, Yastreb used a pulley and cable articulation system. Wide metal rings around the gray nylon canvas undersuit's upper arms served as "anchors" for the upper body articulation system. Yastreb had a regenerative life support system in a rectangular white metal box placed on the chest and abdomen to facilitate movement through Soyuz hatchways. Volynov checked out Khrunov and Yeliseyev's life support and communications systems before returning to the descent module, sealing the hatch, and depressurizing the orbital module. Khrunov went out first, transferring to the Soyuz 4 orbital module while the docked spacecraft were out of radio contact with the Soviet Union over South America. Yeliseyev transferred while the spacecraft were over the Soviet Union. They closed the Soyuz 4 orbital module hatch behind them, then Soyuz 4 Commander Vladimir Shatalov repressurized the orbital module and entered to help Khrunov and Yeliseyev get out of their suits. The spacewalkers delivered newspapers, letters, and telegrams printed after Shatalov lifted off to help prove that the transfer took place. Soyuz 4 and 5 separated after only 4 hr, 35 min together.

Astronautics and Aeronautics, 1969, NASA SP 4014, 1970, p. 12; "Soyuz Spurs Orbiting Space Station Plans," Donald Winston, *Aviation Week & Space Technology*, January 27, 1969, p. 19; *Russian Space History*, Sotheby's Auction Catalog for March 16, 1996 sale of Yastreb undersuit and life support pack, Yastreb blueprints, GTF-2 helmet, and other artifacts, #126-128; *Handbook of Soviet Manned Space Flight*, Nicholas Johnson, Vol. 48, Science and Technology Series, 1980, pp. 151-158.

January 17	**Soyuz 4 landing**
January 18	**Soyuz 5 landing**
March 3	**Apollo 9 launch**

March 6
1969 EVA 2
World EVA 12
U.S. EVA 10

Duration: 0:46
Spacecraft/mission: Apollo 9
Crew: James McDivitt, Russell Schweickart, David Scott
Spacewalkers: Russell Schweickart, David Scott
Purpose: Demonstrate contingency EVA transfer between Apollo LM and CM; test A7L suit and PLSS; test LM and CM ability to support EVA

Lunar Module Pilot (LMP) Russell Schweickart suffered two bouts of vomiting one day after launch and Command Module Pilot (CMP) David Scott reported feeling ill. These symptoms of space motion sickness caused Commander (CDR) James McDivitt and Mission Control to limit Schweickart's scheduled 2-hr EVA to a test of the Apollo A7L Extravehicular Mobility Unit (EMU) and Portable Life Support System (PLSS) inside the Lunar Module (LM) *Spider*'s cabin. For the revised EVA, which would occur in daylight and last less than 1 hr, *Spider* and CM *Gumdrop* would be depressurized and their external hatches opened. The EMU tested on Apollo 9 and used on Apollo lunar missions 11 through 14 weighed about 85 kg (185 lb) fully charged and included three main parts:

* A7L Pressure Garment Assembly (PGA), the "man-shaped" part of the EMU

* Portable Life Support System (PLSS) backpack connected to the PGA by hoses and harnesses

* Oxygen Purge System contingency oxygen supply

The A7L EVA PGA weighed 19.69 kg (43.42 lb). For lunar landing missions the CDR and LMP wore the EVA version both on the lunar surface and during operations requiring them to suit up inside the CM spacecraft. The CMP wore the lighter, stiffer IV version with a thinner integrated thermal meteoroid garment outer layer than the EVA version. On the Apollo 9 through 14 suits, the astronaut entered through a zippered opening running from the front of the crotch up the back. The PGA also included a custom-sized integral boot with a heel clip for securing the legs to the CM couch during launch and reentry. Moonwalkers wore sturdy lunar boots over the integral boots on the lunar surface. During EVA the PGA was worn over the Liquid Cooling Garment (LCG), a nylon-spandex coverall worn next to the astronaut's skin. Cooling water flowed through tiny plastic tubes in the LCG, carrying away excess heat to the sublimator in the PLSS. The PGA pressure bladder was designed to operate at an internal pressure of 25.88 kilopascal (3.75 psi). The neoprene-coated nylon bladder had dipped rubber convolute ("accordion") joints at shoulders, elbows, wrists, hips, knees, and ankles. Cables with reinforced attachment points prevented the convolute joints from ballooning, so they maintained "near-constant volume," and an upper arm bearing with restraining cables improved arm mobility. The chest area of the PGA included inlet and outlet gas connectors and a water connector with inlet and outlet manifolds. The A7L was topped by a locking neck ring to which attached the clear polycarbonate plastic pressure helmet assembly. The Lunar Extravehicular Visor Assembly (LEVA), an assortment of adjustable shades and filters, was worn over the polycarbonate helmet during EVA. The astronaut wore communications carrier assembly "snoopy hats" with redundant microphones and headphones. A drink bag was secured near the astronaut's neck. Custom-sized gloves were attached to wrist rings by rotating quick-disconnect couplings and equipped with a wrist convolute. The EVA gloves included additional outer layers to shield against abrasion from tools and lunar materials. These included high-strength silicone rubber-coated nylon tricot thumb and fingertip shells. The PLSS backpack provided air to the PGA and cooling water to the LCG. The backpack contained the oxygen ventilating circuit, the feedwater and liquid transport loops, the primary oxygen and electrical power subsystems, the EVA communications system, and the remote control unit. A heat exchanger (sublimator) provided cooling; the PLSS vented water vapor as part of the cooling process. Water recharge required less than 10 min. The PLSS provided oxygen and cooling water for about 5 hr of EVA. A hard cover and a thermal blanket covered the assembled unit. A single 16.8V DC battery provided PLSS electricity. The Oxygen Purge System was originally designed in 1967 to provide 30 min of emergency oxygen for breathing and suit cooling. The device could also supplement the main PLSS oxygen supply. A monopole radio communications antenna was considered part of the PLSS, but was mounted on top of the purge system. On mission day 4 Schweickart and McDivitt entered the LM. The astronauts depressurized the two spacecraft and McDivitt opened *Spider*'s hatch. Schweickart was feeling better than expected, so McDivitt allowed him to egress and place his feet in the "golden slipper" foot restraint on *Spider*'s porch. This was the only time the PLSS was tested in space prior to the Apollo 11 Moon landing. Schweickart took photographs while Scott opened the CM hatch, emerged partially, and retrieved thermal exposure samples from *Gumdrop*'s exterior. Scott's SEVA was designed to demonstrate the CMP's ability to prepare the CM for contingency EVA transfer by the LMP and CDR from the LM in the event of docking or IV transfer problems. Scott, who remained connected to *Gumdrop*'s life support system through an umbilical, wore one EVA glove and one IV glove; he found that the IV hand grew slightly warm. Schweickart performed well in the foot restraint, so McDivitt permitted him to test handrails and retrieve thermal samples on *Spider*'s exterior. Movement using the handrails was easier in space than during training, Schweickart reported. He returned to *Spider*, and the astronauts closed the hatches and repressurized the two spacecraft. McDivitt and Schweickart practiced recharging the PLSS before returning to *Gumdrop*.

Chariots for Apollo, NASA SP 4205, Courtney Brooks, *et al*, 1979, pp. 294-298; *Manned Spacecraft Log*, Tim Furniss, 1983, p. 59; "Getting It All Together," George Mueller, in *Apollo Expeditions to the Moon*, NASA SP 350, Edgar Cortwright, editor, 1975, pp. 190, 192; Andrew Chaikin, May 3, 1996; John Charles, May 3, 1996; Joel Powell, May 2, 1996; *Apollo Operations Handbook: Extravehicular Mobility Unit*, Vol. 1, System Description, Rev. IV, MSC-01372-1, June 1968; *Apollo Experience Report - Development of the Extravehicular Mobility Unit*, NASA TN D-8093, Charles Lutz, Harley Stutesman, Maurice Carson, and James McBarron, NASA, November 1975; *Apollo 11 Press Kit*, pp. 117-122; *Apollo 14 Press Kit*, pp. 64-67; interview, David S. F. Portree with James McBarron, July 16, 1996.

March 13   **Apollo 9 splashdown**

May 18-26   **Apollo 10**

July 16   **Apollo 11 launch**

July 20
1969 EVA 3
World EVA 13
U.S. EVA 11
Lunar Surface EVA 1
Duration: 2:32
Spacecraft/mission: Apollo 11
Crew: Neil Armstrong, Edwin Aldrin, Michael Collins
Moonwalkers: Neil Armstrong, Edwin Aldrin
Purpose: Fulfill political requirement of placing a man on the lunar surface; demonstrate ability to perform lunar surface EVA; collect surface samples; deploy EASEP

The first Apollo landing site, in the southern Sea of Tranquility about 20 km (12 mi) southwest of the crater Sabine D, was selected in part because it had been characterized as relatively flat and smooth by the automated Ranger 8 and Surveyor 5 landers, as well as by Lunar Orbiter mapping spacecraft, and therefore unlikely to present major landing or EVA challenges. Two hr after avoiding touchdown in a crater full of 3-m (10-ft) boulders, Armstrong and Aldrin received permission to go outside the LM *Eagle* four and a half hr early. They planned placement of the Early Apollo Scientific Experiment Package (EASEP) and the U.S. flag by studying their landing site through *Eagle*'s twin triangular windows, which gave them a 60-deg field of view. Preparation required longer than the 2 hr scheduled. Armstrong had some initial difficulties squeezing through the hatch with his PLSS. According to veteran moonwalker John Young, a redesign of the LM to incorporate a smaller hatch was not followed by a redesign of the PLSS backpack, so some of the highest heart rates recorded from Apollo astronauts occurred during LM egress and ingress. The Remote Control Unit controls on Armstrong's chest prevented him from seeing his feet. While climbing down the nine-rung ladder, Armstrong pulled the D-ring which deploys the Modular Equipment Stowage Assembly (MESA) folded against *Eagle*'s side and activated the TV camera. Ghostly black and white images of the first lunar EVA were immediately broadcast to at least 600 million people on Earth. After describing the surface ("very fine grained. . . almost like a powder"), Armstrong stepped off *Eagle*'s footpad and into history as the first human to set foot

Apollo 11, 1969 - Edwin Aldrin prepares to join Neil Armstrong (the photographer) on the lunar surface during humanity's first lunar surface EVA. Note the PLSS backpack and lunar overshoes. (AS11-40-5868)

on another world. He reported that moving in the Moon's gravity, one-sixth of Earth's, was "perhaps even easier than the simulations." In addition to fulfilling President John F. Kennedy's mandate to land a man on the Moon before the end of the 1960s, Apollo 11 was an engineering test of the Apollo system; therefore, Armstrong snapped photos of the LM so engineers would be able to judge its post-landing condition. He then collected a contingency soil sample using a sample bag on a stick. He folded the bag and tucked it into a pocket on his right thigh. He removed the TV camera from the MESA, made a panoramic sweep, and mounted it on a tripod 12 m (40 ft) from the LM. The TV camera cable remained partly coiled and presented a tripping hazard throughout the EVA. Aldrin joined him on the surface and tested methods for moving around, including two-footed kangaroo hops. The PLSS backpack created a tendency to tip backwards, but neither astronaut had serious problems maintaining balance. Loping became the preferred method of movement. The astronauts reported that they needed to plan their movements six or seven steps ahead. The fine soil was quite slippery. Aldrin remarked that moving from sunlight into *Eagle*'s shadow produced no temperature change inside the suit, though the helmet was warmer in sunlight, so he felt cooler in shadow. Together the astronauts planted the U.S. flag - the ground was too hard to permit them to insert the pole more than about 20 cm (8 in) - then took a phone call from President Richard Nixon. The MESA failed to provide a stable work platform and was in shadow, slowing work somewhat. As they worked, the moonwalkers kicked up gray dust which soiled the outer part of their suits, the integrated thermal meteoroid garment. They deployed the EASEP, which included a passive seismograph and a laser ranging retroreflector. Then Armstrong loped about 120 m (400 ft) from the LM to snap photos at the rim of East Crater while Aldrin collected two core tubes. He used the geological hammer to pound in the tubes - the only time the hammer was used on Apollo 11. The astronauts then collected rock samples using scoops and tongs on extension handles. Many of the surface activities took longer than expected, so they had to stop documented sample collection halfway through the allotted 34 min. During this period Mission Control used a coded phrase to warn Armstrong that his metabolic rates were high and that he should slow down. He was moving rapidly from task to task as time ran out. Rates remained generally lower than expected for both astronauts throughout the walk, however, so Mission Control granted the astronauts a 15-min extension. Aldrin entered *Eagle* first. With some difficulty the astronauts lifted film and two sample boxes containing more than 22 kg (48 lb) of lunar surface material to the LM hatch using a flat cable pulley device called the Lunar Equipment Conveyor. Armstrong then jumped to the ladder's third rung and climbed into the LM. After transferring to LM life support, the explorers lightened the ascent stage for return to lunar orbit by tossing out their PLSS backpacks, lunar overshoes, one Hasselblad camera, and other equipment. Then they lifted off in *Eagle*'s ascent stage to rejoin CMP Michael Collins aboard the CM *Columbia* in lunar orbit.

Apollo 11 mission transcript, p. 374-376, 377, 379, 394; *Apollo Lunar Surface Journal*, Eric Jones (http://www.hq.nasa.gov/office/pao/History/alsj/); *Chariots for Apollo*, NASA SP 4205, Courtney Brooks, *et al*, 1979, pp. 346-349; *To a Rocky Moon*, Donald Wilhelms, 1993, p. 201-205; *Astronautics and Aeronautics, 1969*, pp. 217-220; *Apollo 11 Mission Report*, NASA SP-238, 1971, pp. 21-25, 129-131; "Mobility Unhindered by Bulky Space Suit," Warren Wetmore, *Aviation Week & Space Technology*, July 28, 1969, p. 35; "Astronauts Detail Lunar Flight Experience," *Aviation Week & Space Technology*, August 18, 1969, p. 19.

July 24	Apollo 11 splashdown
October 11-16	Soyuz 6
October 12-17	Soyuz 7
October 13-18	Soyuz 8

November 19

1969 EVA 4
World EVA 14
U.S. EVA 12
Lunar Surface EVA 2
Duration: 3:39
Spacecraft/mission: Apollo 12
Crew: Charles Conrad, Alan Bean, Richard Gordon
Moonwalkers: Charles Conrad, Alan Bean
Purpose: Deploy ALSEP; collect samples

Conrad guided LM *Intrepid* over Surveyor Crater and lost sight of the surface at an altitude of 12 m (40 ft) because of dust kicked up by the descent engine, but still managed to land 180 m (600 ft) from Surveyor 3, the primary target of EVA 2. Landing near the old robot amply demonstrated the pinpoint landing capability of the Apollo system. This was critical for planning EVA traverses for future Apollo missions. The first EVA of Apollo 12 started about 50 min late. Conrad opened the MESA while climbing down the ladder, activating the color TV camera. Conrad then dropped onto *Intrepid*'s forward footpad from the ladder's last rung, calling out, "Whoopie! Man, that may have been a small one for Neil, but that's a long one for me!" Even before Bean joined him on the surface Conrad reported that he was getting dirty from lunar dust. Bean removed the camera from the MESA and put it on a tripod, in the process pointing it at the Sun. The camera's vidicon tube was damaged, ending EVA TV for Apollo 12. Conrad and Bean planted the U.S. flag and collected rocks near the LM, then deployed the Advanced Lunar Science Experiment Package (ALSEP) north of Surveyor Crater, about 180 m (600 ft) from *Intrepid*. Dust kicked up by the astronauts stuck to the ALSEP instruments. The dust was not slippery, though Conrad took a harmless spill because of uneven footing. The astronauts had some difficulty removing from *Intrepid*'s side the plutonium fuel cartridge for ALSEP's SNAP-27 nuclear power source. The cartridge was hot enough to melt a hole in a space suit. The astronauts saw *Yankee Clipper*, piloted by CMP Richard Gordon, as a bright star passing overhead. Conrad and Bean attempted to dust each other off before climbing back into *Intrepid* for the night.

Apollo 12 Preliminary Science Report, NASA SP 235, 1970, pp. 31-33; *Apollo Lunar Surface Journal*, Eric Jones, 1995 (http://www.hq.nasa.gov/office/pao/History/alsj/); *Astronautics and Aeronautics 1969*, pp. 376-377; "Exuberance Sets Tone of First EVA," *Aviation Week & Space Technology*, November 24, 1969, pp. 19-21; *A Man on the Moon*, Andrew Chaikin, 1993, p. 267.

November 19

1969 EVA 5
World EVA 15
U.S. EVA 13
Lunar Surface EVA 3
Duration: 3:48
Spacecraft/mission: Apollo 12
Crew: Charles Conrad, Alan Bean, Richard Gordon
Moonwalkers: Charles Conrad, Alan Bean
Purpose: Perform traverse to Surveyor 3 and landmark craters; retrieve pieces of Surveyor 3; collect samples

On this EVA, Conrad and Bean became the first to undertake a long (1800-m, or 6000-ft) lunar traverse. Objects on the surface appeared closer than they really were, and the bottoms of craters were hidden in shadow. In their section of the *Apollo 12 Preliminary Science Report*, Conrad and Bean describe the effects on lunar surface mobility of lunar gravity and the Apollo EMU used for Apollos 11, 12, and 14:

> On the Moon, the. . . legs never seem to get tired. The problem with the suit is that it does not always bend as the wearer wants to bend. For example, the suit bends fairly well in the knees and ankles, but it does not want to bend near the top of the thigh. This. . . results in loping in a stiff-legged fashion - running with straight legs, landing flat-footed, and then pushing off with the toes.

The traverse had three major objectives - site exploration, geological sample collection, and Surveyor 3 inspection. Conrad and Bean first cut the TV camera damaged during the first EVA from its cable and made room for it in the EVA rock box used during the first EVA so it could be returned to Earth for examination. They moved to the ALSEP site, then to Bench Crater, southwest of the LM. At Sharp Crater they collected a core and a trench sample, then they moved to Halo Crater, south of the LM, where they collected two cores. Along the way they saw spatters of cooled molten glass on rocks and occasional brightly colored rocks. They also noted that the surface was gray or brown depending on Sun angle and direction. Two hr into the EVA they received a "go" for a 4-hr EVA. The astronauts stated later that they felt little fatigue and could have continued for twice as long as the three and half hr originally scheduled. They skirted the south rim of Surveyor Crater to the southeast. Surveyor 3 sat on a 12-deg slope, 45 m (150 ft) inside the crater. The moonwalkers were puzzled by the apparent change in Surveyor's white color caused by a fine coating of tan lunar dust. It looked brown through the EMU's gold-coated visor. Conrad and Bean then became the first extraterrestrial archaeologists, collecting pieces of Surveyor 3 for study by scientists interested in the long-term of effects on equipment of lunar surface conditions. Avoiding residual propellant and sharp edges, the astronauts used a cutting tool ("bolt cutter") to remove Surveyor 3's TV camera, sample scoop, and samples of wire, glass, and metal. The astronauts reported that Surveyor 3 differed from the mockup they used in training. Then they moved back to the LM, stowed the rock boxes, and closed out EVA 2. The astronauts reported that fighting the internal pressure of their gloves made their hands tired. Bean asked CapCom Ed Gibson to tell Fred Haise, Apollo 13 LMP, to do hand exercises to prepare for his visit to the Moon. Gibson used the code phrase "We'd like an EMU check" to warn the astronauts to slow down after their heart rates reached about 160 beats/min. The astronauts reported that their integrated thermal meteoroid garments were severely worn by lunar dust abrasion. Dust tracked into *Intrepid* became weightless during ascent to lunar orbit, making breathing difficult without their helmets. During the flight home in *Yankee Clipper*, the astronauts had to clean lunar dust off the air filter screens every 2 to 3 hr. After splashdown on November 24, the astronauts had difficulty telling themselves apart in their more than 500 photos.

Catalog of Apollo Experiment Operations, NASA RP 1317, Thomas Sullivan, January 1994, pp. 129-131; *Apollo 12 Preliminary Science Report*, NASA SP 235, 1970, pp. 33-37; *Astronautics and Aeronautics 1969*, pp. 376-377; *Apollo Lunar Surface Journal*, Eric Jones, 1995 (http://www.hq.nasa.gov/office/pao/History/alsj/); "Apollo Yields Rich Lunar Return," Zack Strickland, *Aviation Week & Space Technology*, November 24, 1969, pp. 16-18; "Astronauts Urge Longer-Duration EVAs," Zack Strickland, *Aviation Week & Space Technology*, December 1, 1969, pp. 17-20; *A Man on the Moon*, Andrew Chaikin, 1994, pp. 266-267, 276.

November 24   **Apollo 12 splashdown**

1970

April 11-17	Apollo 13
June 2-19	Soyuz 9

1971

January 31	Apollo 14 launch

February 5
1971 EVA 1
World EVA 16
U.S. EVA 14
Lunar Surface EVA 4
Duration: 4:49
Spacecraft/mission: Apollo 14
Crew: Alan Shepard, Edgar Mitchell, Stuart Roosa
Moonwalkers: Alan Shepard, Edgar Mitchell
Purpose: ALSEP deployment

Early on this date the LM *Antares* touched down at Fra Mauro, which was originally the landing site of James Lovell and Fred Haise on the aborted Apollo 13 mission. Mitchell described the landscape outside *Antares'* viewports as "choppy, undulating," and added that "I can see several ridges and rolling hills of perhaps 35 to 40 ft [10.5 to 12 m] in height." Communications problems delayed the first EVA's start by 49 minutes. Shepard stepped onto the Moon and described the surface as "so soft that it comes all the way to the top of the [LM] footpads; it's even folded over the sides to some degree. . ." Looking ahead to the second EVA, which would include a climb to the rim of Cone Crater, he added that "it looks as though we have a good traverse route up to the top of the Cone." After Mitchell joined Shepard on the surface, the astronauts collected a 19.5-kg (42.9-lb) contingency sample, then deployed the TV camera (taking care not to point it at the Sun), S-band dish antenna, and U.S. flag. Shepard's suit had red stripes at the knees and shoulders so he could be identified in photographs after the mission. His helmet also bore a red stripe. Apollo 13 Commander James Lovell also had identifying stripes on his suit. A hinged center shade section was added to the LEVA, providing additional eye protection when the astronaut walked toward the Sun under low Sun-angle conditions. The astronauts deployed the ALSEP experiments about 150 m (495 ft) west of *Antares*, then set up the laser ranging retroreflector 30 m (100 ft) beyond that. The EVA was extended by 30 min to partly compensate for the late start. In all, the astronauts covered about 550 m (1815 ft) before returning to *Antares* to eat and rest. Meanwhile, in lunar orbit, CMP Stuart Roosa conducted photography to aid in selection of the remaining Apollo sites and surface EVA planning.

Apollo 14 Preliminary Science Report, NASA SP 272, 1971, pp. 34-35; *Apollo Lunar Surface Journal*, Eric Jones, 1995 (http://www.hq.nasa.gov/office/pao/History/alsj/); *Astronautics and Aeronautics 1971*, NASA SP 4016, pp. 27, 41; *To a Rocky Moon*, Donald Wilhelms, 1993, pp. 250.

February 6
1971 EVA 2
World EVA 17
U.S. EVA 15

Lunar Surface EVA 5
Duration: 4:46
Spacecraft/mission: Apollo 14
Crew: Alan Shepard, Edgar Mitchell, Stuart Roosa
Moonwalkers: Alan Shepard, Edgar Mitchell
Purpose: Geological traverse to rim of Cone Crater

Apollo 14 marked a modest upgrade in lunar surface EVA capabilities. The EMUs Shepard and Mitchell donned were modified to include the Buddy Life Support System (BLSS), a 2.5-m (8-ft) umbilical which allowed an astronaut with a malfunctioning PLSS to draw cooling water from his companion's healthy PLSS until he could return to the LM. For this EVA the BLSS umbilical was stowed on the NASA MSC-built Modularized Equipment Transporter (MET), a 9-kg (19.8-lb), two-wheeled "rickshaw" cart for hauling tools, photographic equipment, and sample containers. The MET had tires inflated to very low pressure which ballooned in lunar vacuum. The astronauts left *Antares* two and a half hr earlier than scheduled. The EVA was planned as a 3-km (1.8-mi) traverse - probably the longest practical for an astronaut wearing an Apollo EMU if useful work was to be performed along the way and a safety margin maintained. Their objective was the rim of Cone Crater, a meteoroid impact pit 300 m (1000 ft) wide. Geologists saw it as a natural excavation laying bare eons of lunar geological history. Reaching the elevated rim was considered important because it should have contained the oldest rocks. Shepard and Mitchell headed east toward the crater, taking turns towing the MET. They began climbing the blanket of loose debris around Cone 90 min into the EVA, about 850 m (2800 ft) from the crater rim. The MET had a tendency to tip and became difficult to pull as the number of boulders increased. Shepard and Mitchell finally resorted to carrying it. Mitchell noted problems with dust, saying that, "we're filthy as pigs. . . everything's going to be covered with dust before long." Shepard's heart rate climbed to 150 beats per min and Mitchell's right EMU wrist cable broke, impeding his hand movements. Shepard said after the EVA, however, that the worst problem was "the undulating terrain where you simply couldn't see more than 100 to 150 yards [90 to 135 m] away from you. Consequently, you were never quite sure what landmark would appear when you topped the next ridge. We were very surprised when we. . . approached the ridge which we thought to be the rim of Cone Crater, to find there was another one behind it. . . I think if we had wanted to reach the top of the crater and did nothing else, that we could have done that within the time period allotted." The astronauts received a 30-min extension, but were finally compelled to abandon their quest for the rim. They obtained only one of three planned core tubes, 16 photographs, and 10 kg (22 lb) of samples during their Cone traverse. Despite their problems, the West German newspaper *Frankfurter Rundschau* reported that what the Apollo 14 astronauts achieved "couldn't have been done by a Lunokhod," referring to the Soviet Union's Lunokhod 1 teleoperated robot, which was exploring Mare Imbrium at this time. In all the astronauts collected more than 43 kg (94.6 lb) of samples.

Apollo 14 Preliminary Science Report, NASA SP 272, 1971, p. 36; *Apollo Lunar Surface Journal*, Eric Jones, 1995 (http://www.hq.nasa.gov/office/pao/History/alsj/); *Astronautics and Aeronautics 1971*, NASA SP 4016, 1972, pp. 27, 41, 43; *Astronautics and Aeronautics 1970*, NASA SP 4015, 1972, p. 338; *Roundup*, MSC, December 4, 1970, p. 1; *A Man on the Moon*, Andrew Chaikin, 1994, pp. 369-375; *To a Rocky Moon*, Donald Wilhelms, 1993, pp. 250-255.

February 9	Apollo 14 splashdown
April 23-24	Soyuz 10/Salyut 1
June 6-29	Soyuz 11/Salyut 1

July 30
1971 EVA 3
World EVA 18
U.S. EVA 16
Lunar Surface EVA 6
Duration: 0:33
Spacecraft/mission: Apollo 15
Crew: David Scott, James Irwin, Alfred Worden
Moonwalker: David Scott
Purpose: Survey Hadley-Apennine landing site from top hatch of LM *Falcon*

This SEVA was partly a response to the surface navigation problems experienced on Apollo 14. Commander David Scott stood atop *Falcon*'s ascent engine cover with his shoulders through the top hatch so he could get his bearings at the complex Hadley-Apennine landing site ahead of the surface EVAs. He told Edgar Mitchell in Mission Control that the site was "exactly like what you had on Apollo 14. It's very hummocky, and, as you know, in this kind of terrain you can hardly see over your eyebrows." Scott reported that 5000-m (16,500-ft) Mt. Hadley glowed gold and brown in the lunar morning sunlight, and that there were no large boulders to hinder the progress of the Lunar Roving Vehicle (LRV) during the next day's scheduled traverse south to St. George Crater.

Astronautics and Aeronautics 1971, NASA SP 4016, p. 203-204; *To a Rocky Moon*, Donald Wilhelms, 1993, p. 270-271; *A Man on the Moon*, Andrew Chaikin, 1994, pp. 414-415.

July 31
1971 EVA 4
World EVA 19
U.S. EVA 17
Lunar Surface EVA 7
Duration: 6:34
Spacecraft/mission: Apollo 15
Crew: David Scott, James Irwin, Alfred Worden
Moonwalkers: David Scott, James Irwin
Purpose: Deploy LRV; traverse south to St. George Crater; deploy ALSEP

Unlike the Apollo 11, 12, and 14 astronauts, Scott and Irwin waited to charge their PLSS back-packs with water until they reached the surface. *Falcon* landed with one footpad in a small crater, causing it to tilt. Scott estimated that, because of *Falcon*'s cant, the backpacks tilted 30 deg during recharge. This prevented them from charging fully and created bubbles in Irwin's PLSS which set off false failure warnings throughout the first EVA. With improved A7LB space suits for longer EVAs, the LRV for greater mobility, and a complex landing site outside the equatorial "Apollo zone," Apollo 15 represented the beginning of a new era of lunar exploration. The improved EMU was more comfortable and retained provisions for the BLSS umbilical, which was stowed on the LRV for Apollos 15, 16, and 17. Added consumables meant that an astronaut could manage a 7-hr EVA and a 5-mi "walkback" if the LRV failed. New convolute joints permitted kneeling, though with some difficulty. Scott set up the improved TV camera while Irwin collected the contingency sample. The astronauts tended to move using two-footed kangaroo hops rather than slow-motion loping. Unstowing and deploying the LRV took longer than the 20 min allotted. Though it was planned as a one-man task, under lunar surface conditions unfolding the vehicle required both astronauts. The Boeing-built LRV was one of the most important additions to Apollo's capabilities. The LRV System included the rover, a structure supporting it within LM descent stage Quadrant 1, and a deployment mechanism. The LRV's empty weight was 205 kg (455 lb); loaded with two astronauts in EMUs, equipment, and lunar samples, it weighed up to 691 kg (1535 lb). The vehicle had a 35-cm (14-in) ground clearance and a 2.25-m (90-in) wheel-base, and was 3.05 m (122 in) long. Top speed was about 21 km (13 mi)/hr. Two independent sets of 36V silver-zinc batteries provided sufficient power for a 62.4-km (39-mile) traverse at 16 km (10 mi)/hr. One set was sufficient for operation. The batteries were located on the forward chassis between the front wheels. Each of the four wheels had a separate traction drive motor and was independently steerable. If a drive motor failed, the affected wheel freewheeled. The LRV had Ackerman steering - that is, it could move sideways if necessary, and could turn within its own length. Conventional steering was also available. Each wheel had a mesh wire tire with a metal chevron frame and inner frame. A T-bar handle between the seats controlled steering, braking, and acceleration. An astronaut braked the LRV by sliding the handle backward; accelerated forward by bending the handle forward; accelerated reverse by bending the handle backward; and turned the rover by bending the handle in the appropriate direction. A "dashboard" control console ahead of the T-bar displayed speed, pitch, distance traveled, bearing to last point of initialization, and distance from the LM. The rear chassis carried tools and rock boxes on EVAs. Seatbelts held the astronauts in their seats against the LRV's rolling and bouncing. A color television camera mounted on the front of the LRV could be pointed and zoomed by a controller in Houston when the LRV's high-gain antenna was aimed at the Earth. This allowed scientists to investigate traverse stops independently. Irwin took the first of several harmless Apollo 15 falls during LRV deployment. The astronauts found that the rover had no front-wheel steering, but Scott was able to maneuver the vehicle using only rear steering. They also found that their seatbelts barely fit around their pressurized EMUs. About 3 hr into the EVA Scott and Irwin set out on a 10.3-km (6.2-mi) traverse south along the rim of Elbow Crater to 2.25-km-dia (1.4-mi-dia) St. George Crater, near Hadley Rille. During the traverse the astronauts reduced suit cooling to avoid becoming cold while their metabolic rates were low. They had some difficulty with "zero phase lighting" (light reflected from the landscape opposite the Sun) which made obstacles difficult to discern. The astronauts used a rake to collect "walnut-sized samples" near St. George Crater. Flight controllers in Houston operated the LRV camera so geologists on Earth could explore the lunar landscape telerobotically and guide the astronauts in collecting samples. The astronauts then drove back to *Falcon* to deploy the ALSEP. They located the central station 110 m (335 ft) west of the LM and drilled a hole in the ground for the heatflow experiment probe. Scott used more oxygen than expected, so flight controllers terminated the EVA 30 min early and considered cutting back the second EVA. Dust made PLSS connectors tight and difficult to operate. After the EVA Irwin was extremely thirsty because his drink bag failed to operate (it refused to supply water during all of the Apollo 15 EVAs). In addition, both astronauts suffered pain in their fingers caused by their fingernails pressing hard against their glove fingertips. Irwin needed help to remove his gloves, and elected to trim his nails before the second EVA. Scott left his fingernails as they were to avoid reducing his dexterity.

Apollo 15 Preliminary Science Report, NASA SP 289, 1972, pp. 1.5-1.6; *Apollo Lunar Surface Journal*, Eric Jones, 1995 (http://www.hq.nasa.gov/office/pao/History/alsj/); "Added Mobility Spurs Lunar Harvest," Zack Strickland, *Aviation Week & Space Technology*, August 9, 1971, pp. 13-17; *Astronautics and Aeronautics 1971*, NASA SP 4016, p. 204-205; *A Man on the Moon*, Andrew Chaikin, 1994, pp. 422-424; *To a Rocky Moon*, Donald Wilhelms, 1993, pp. 272-274; *Catalog of Apollo Experiment Operations*, Thomas Sullivan, NASA RP 1317, January 1994, pp. 79-83; *Lunar Roving Vehicle Operations Handbook*, Boeing LRV Systems Engineering, Contract NAS8-25145, November 6, 1972, pp. 1.1-1.21, 6.2-6.6; interview, David S. F. Portree with Joseph P. Loftus, Jr., May 30, 1996.

August 1

1971 EVA 5
World EVA 20
U.S. EVA 18
Lunar Surface EVA 8
Duration: 7:13
Spacecraft/mission: Apollo 15
Crew: David Scott, James Irwin, Alfred Worden
Moonwalkers: David Scott, James Irwin
Purpose: Geological traverse southeast to Mt. Hadley Delta; deploy heat flow experiment and U.S. flag

Based on their experience during the first traverse, mission planners modified the second EVA to maximize time spent doing science and minimize driving. The start of the second EVA was delayed 30 min by PLSS backpack recharge difficulties. When they did reach the surface, Scott and Irwin were delighted to discover that LRV front steering had become operational. They began a 12.5-km (7.5-mi) traverse southeast to the foot of the Hadley Delta mountain and back, passing Index, Arbeit, Crescent, Spur, and Window craters. The LRV climbed Hadley Delta's slopes at 10 kph (6 mph) with no difficulty. Scott and Irwin climbed 300 ft above *Falcon*, which was more than 5 km (3 mi) away. Soft material on the slopes provided poor footing, and the LRV began to slide while parked. Irwin held the rover while Scott hopped off to collect a green crystalline rock. At Spur Crater they collected the "Genesis Rock," which today is still believed to be a piece of original lunar crust more than four billion yr old. Scott called Spur a "gold mine" of interesting geological samples, so their time there was extended to 49 min. CapCom Joe Allen assisted the astronauts via the LRV camera by warning Irwin that he was about to lose his sample bag. The elevation clutch on the LRV camera began to slip. Irwin's vertical PLSS antenna snapped off, and Scott taped it on horizontally. They collected so many samples at Spur that the LRV bounced when they dropped the rock box on it. They then had to rush because they were approaching the "walkback" limit of their EMUs. In contrast to the first EVA, Scott used about as much oxygen as expected. The astronauts found navigating back to the LM difficult until they encountered their own outbound tracks. Back at *Falcon*, Scott drilled a core hole, encountering much resistance and hurting his hands. Then the 3-m-long (10-ft-long) core tube could not be removed. On advice from Mission Control, the astronauts abandoned the tube until the next EVA. They planted the U.S. flag at the EVA's end.

"Added Mobility Spurs Lunar Harvest," Zack Strickland, *Aviation Week & Space Technology*, August 9, 1971, pp. 13-17; *Apollo 15 Preliminary Science Report*, NASA SP 289, 1972, pp. 1.5-1.6; *Apollo Lunar Surface Journal*, Eric Jones, 1995 (http://www.hq.nasa.gov/office/pao/History/alsj/); *To a Rocky Moon*, Donald Wilhelms, 1993, pp. 275-277; *A Man on the Moon*, Andrew Chaikin, 1994, p. 426, 429.

August 2

1971 EVA 6
World EVA 21
U.S. EVA 19
Lunar Surface EVA 9
Duration: 4:20
Spacecraft/mission: Apollo 15
Crew: David Scott, James Irwin, Alfred Worden
Moonwalkers: David Scott, James Irwin
Purpose: Geological traverse west to Scarp crater and Hadley Rille

The astronauts propped their PLSS backpacks upright for charging, eliminating their bubble problems. With this EVA, the fifth of his career, Scott became the record-holder for number of career EVAs. His record was not beaten until 1984, when cosmonauts Leonid Kizim and Vladimir Solovyov performed six EVAs outside Salyut 7. The EVA started 1 hr, 45 min late to let the crew rest after they experienced irregular heartbeats. This was traced later to potassium deficiency, complicated in Irwin's case by failure of his drink bag. The EVA was shortened to protect *Falcon*'s planned liftoff time. Scott and Irwin managed to free the core tube which became stuck on EVA 2, but could not take it apart to stow it because the LRV vise was assembled backwards on Earth. They used a wrench and lost 28 min. Irwin and Scott started their traverse 1 hr, 20 min into the EVA. They drove 5.1 km (3 mi) west to Scarp Crater, then turned northwest to Hadley Rille, with stops at Rim Crater and a feature called The Terrace. This EVA marked the first time Apollo astronauts passed out of sight of their LM. After Scott and Irwin returned to *Falcon*, Apollo 15 Flight Director Gerald Griffin called the LRV a "great little machine," and added that, "I hate to leave it up there." During the three traverses, the LRV was used to collect nearly 80 kg of samples and covered almost 50 km (28 mi). According to scientist-astronaut Harrison Schmitt, in training as LMP on Apollo 17, ". . . we had a fantastic exploration mission. There's just no question in my mind that we sent two very competent observational scientists to the Moon." *The New York Times* on this date pointed to the obvious success of Apollo 15 and reminded its readers that the Apollo program would end with Apollo 17. The paper lamented how a "vast and complex technology developed at the cost of billions of dollars over the last decade is being abandoned even as its vast potentialities are being demonstrated."

"Added Mobility Spurs Lunar Harvest," Zack Strickland, *Aviation Week & Space Technology*, August 9, 1971, pp. 13-17; *Astronautics and Aeronautics 1971*, NASA SP 4016, pp. 205-206, 216, 217; *Apollo 15 Preliminary Science Report*, NASA SP 289, 1972, pp. 1.5-1.6; *Apollo Lunar Surface Journal*, Eric Jones, 1995 (http://www.hq.nasa.gov/office/pao/History/alsj/); *A Man on the Moon*, Andrew Chaikin, 1994, pp. 439-440; *To a Rocky Moon*, Donald Wilhelms, 1993, pp. 278.

August 5
1971 EVA 7
World EVA 22
U.S. EVA 20
Deep Space EVA 1
Duration: 0:41
Spacecraft/mission: Apollo 15
Crew: David Scott, James Irwin, Alfred Worden
Spacewalker: Alfred Worden
Purpose: Retrieve film from SIM bay of CM *Endeavour*

Alfred Worden became the first astronaut to "go EVA" beyond the protective envelope of Earth's inner magnetosphere. The planned 1-hr EVA, 273,600 km (171,000 mi) from Earth, was televised via a camera on a boom extended from the CM hatch. Using handrails and foot restraints, he had no difficulty making three round trips to the Scientific Instrument Module (SIM) bay built into the side of *Endeavour*'s Service Module (SM). Irwin guided Worden's 8.3-m (27.4-ft) tether from CM hatch. Worden first retrieved the 39-kg (86-lb) Itek panoramic camera cassette, which he tethered to his arm and carried to Irwin at the hatch. Though Worden's metabolic rates remained acceptable throughout the EVA, CapCom Karl Henize warned him not to rush. On the second trip, he removed the 10-kg (22-lb) cassette from the Fairchild mapping camera. Worden made an unplanned third trip to inspect SIM bay instruments which had malfunctioned.

"Worden Takes First Deep Space Walks," *Aviation Week & Space Technology*, August 9, 1971, pp. 22-23.

1972

April 16   Apollo 16 launch

April 21
1972 EVA 1
World EVA 23
U.S. EVA 21
Lunar Surface EVA 10
Duration: 7:11
Spacecraft/mission: Apollo 16
Crew: John Young, Charles Duke, Thomas Mattingly
Moonwalkers: John Young, Charles Duke
Purpose: Deploy ALSEP; deploy LRV; geological traverse to Flag crater

Apollo 16, 1972 - The LRV is clearly visible in this photograph of John Young taken by Charles Duke at Apollo 16's Descartes landing site. (AS16-117-18825)

Apollo 16 was the only expedition planned to the lunar highlands. LM *Orion* landed several hours late because of a malfunction in CM *Casper*'s Service Propulsion System. For a time Mission Control feared that the malfunction might prevent the CM from making rendezvous with the LM ascent stage after liftoff from the lunar surface. Mission Control delayed the first EVA's start until after an 8-hr rest period, during which Commander Young and LMP Duke studied the Descartes site through *Orion*'s windows. They reported that Descartes was rockier and hillier than previous sites. Duke had trouble getting into his EMU because he had grown 4 cm (1.5 in) in weightlessness. This physiological effect of weightlessness was not taken into account during suit fitting in the Apollo program. A problem with *Orion*'s steerable antenna delayed EVA start by 1 hr and prevented Young's first steps on the Moon from being televised. For the first time the astronauts collected no contingency sample. They deployed the U.S. flag, then began ALSEP deployment about 90 min into the EVA. Duke used an improved drill to collect a 2.6-m (8.6-ft) core. He inserted the heat flow probe, which was linked to the ALSEP central station by a cable. Young then accidentally walked over the cable, tearing it loose from the central station. Mission Control began study of a possible repair during the second EVA. The deployment lanyard on the cosmic ray experiment broke, leaving the crew uncertain as to whether the instrument was fully deployed. They then deployed the LRV from *Orion*'s side. The rover had no rear steering and one of its batteries read low. However, rear steering returned and the battery read normal 40 min into the traverse. Young and Duke drove past Flag, Spook, Buster, and Plum craters. Near Plum they collected an 11.7-kg (25.7-lb) rock through a "videoconference" with Earth using the LRV camera. The LRV bounced a great deal during the traverses. Following the traverse, Duke operated a movie camera while Young performed LRV traction tests known jocularly as the "Grand Prix." When Mission Control relayed the news that the House of Representatives approved FY1973 Space Shuttle funding the day before, John Young leaped 1 m (3 ft) and saluted the flag. (Young went on to command the first Shuttle mission, STS-1, in 1981.) Duke jumped too, but slipped and fell on his PLSS. After the EVA the astronauts reported the usual dust problems - stuck zippers and glove disconnects and indicators scratched and difficult to read.

Astronautics and Aeronautics 1972, NASA SP 4017, pp. 146, 155; *Manned Spaceflight Log*, Tim Furniss, 1983, p. 82; *Apollo Lunar Surface Journal*, Eric Jones, 1995 (http://www.hq.nasa.gov/office/pao/History/alsj/); *Apollo 16 Preliminary Science Report*, NASA SP 315, 1972, pp. 2.4-2.5, 6.6-6.7; *To a Rocky Moon*, Donald Wilhelms, 1993, pp. 295, 318; interview, David S. F. Portree with John Young, June 13, 1996.

April 22
1972 EVA 2
World EVA 24
U.S. EVA 22
Lunar Surface EVA 11
Duration: 7:23
Spacecraft/mission: Apollo 16
Crew: John Young, Charles Duke, Thomas Mattingly
Moonwalkers: John Young, Charles Duke
Purpose: Geological traverse to Stone Mountain

Prior to this second Apollo 16 EVA, Mission Control vetoed repair of the heat flow cable because it would take too much time and possibly short-circuit the ALSEP central station. Experimenters announced that they would strengthen the cable for the Apollo 17 mission. During egress a 5-cm portion of Young's PLSS antenna broke off against the LM's hatch frame, causing a small drop in signal strength. The astronauts moved the cosmic ray experiment to an LM footpad out of the Sun because it showed signs of overheating. Young noticed that he felt cooler in *Orion*'s shadow, where the surface temperature was minus 84 deg C (minus 120 deg F). During the Apollo 16 EVAs surface temperature in the Sun was 88 deg C (190 deg F). Young and Duke then mounted the LRV and climbed Survey Ridge to Stone Mountain. A problem with LRV steering and traction was found to be caused by "mismatched power modes," and was solved by changing switch settings on the "dash board" mounted ahead of the control T-handle. The astronauts collected core samples on Stone Mountain and reported that the view of their landing site was "just dazzling." The LM was barely visible in the distance. During descent to *Orion* the pitch meter on the dashboard pegged at 20 deg of slope. The astronauts collected a thin layer of surface material using adhesive plates. As with most Apollo lunar surface EVAs, some activities and stops were deleted because of insufficient time.

Astronautics and Aeronautics 1972, NASA SP 4017, pp. 146-147; *Apollo 16 Preliminary Science Report*, NASA SP 315, 1972, pp. 2.4-2.5, 5.2-5.3, 6.6-6.7; *Apollo Lunar Surface Journal*, Eric Jones, 1995 (http://www.hq.nasa.gov/office/pao/History/alsj/); *To a Rocky Moon*, Donald Wilhelms, 1993, p. 298; interview, David S. F. Portree with John Young, June 13, 1996.

April 23
1972 EVA 3
World EVA 25
U.S. EVA 23
Lunar Surface EVA 12
Duration: 5:40
Spacecraft/mission: Apollo 16
Crew: John Young, Charles Duke, Thomas Mattingly
Moonwalkers: John Young, Charles Duke
Purpose: Traverse to Smoky Mountain

The landing delay on April 21 caused more water than expected to be used in cooling *Orion*'s avionics. Because the LM's cooling water supply was running low, consideration was given to deleting this third Apollo 16 EVA, a move the Apollo science "back room" at Mission Control vehemently opposed. The last Apollo 16 EVA began 30 min early, but overall length was cut by two hr and the traverse was shortened by five stops. Young and Duke drove to the foot of Smoky Mountain, near North Ray Crater, where they spent 1 hr, 20 min sampling and taking magnetic field readings using the Lunar Portable Magnetometer. They found the largest remnant magnetic field discovered on the Moon. The astronauts commented on the thick dust they kicked up and the generally shattered appearance of the area. Geologists in the back room asked Young to look at North Ray's bottom, but he turned down the request, saying that, "That rascally rim slopes about 10 or 15 degrees. . . then all of a sudden . . . I've got to go 100 yards [92 m] down a 25 to 30 degree slope and I don't think I'd better." The astronauts collected samples off 10-m-high (33-ft-high), 20-m-long (66-ft-long) House Rock, the largest boulder sampled during Apollo. Young found that by hopping into the air and landing on his feet, the weight of his suit overcame the suit's internal pressure, so he could get to his knees and pick up rocks without using geological tools. The LRV suffered temporary navigational computer failure, but Young and Duke knew where they were from the Sun's position. They trended back toward their outbound tracks so they could follow them back to the LM, but spotted their spacecraft before they found their tracks. The LRV reached its highest speed on the Moon - 22 kph (13 mph) - rolling down a 15-deg slope during return to *Orion*. Young left the LRV parked 50 m (164 ft) east of the LM, then helped Duke load 96.6 kg (212.5 lb) of lunar samples into the spacecraft. Controllers on Earth used the LRV camera to track *Orion*'s ascent stage as it left Descartes behind. The LM's avionics cooling water ran out as it completed docking with the CM *Casper*.

Astronautics and Aeronautics 1972, NASA SP 4017, p. 148; *Apollo 16 Preliminary Science Report*, NASA SP 315, 1972, pp. 2.4-2.5, 5.2-5.3, 6.6-6.7; *To a Rocky Moon*, Donald Wilhelms, pp. 300-301; *A Man on the Moon*, Andrew Chaikin, 1994, pp. 488-489; interview, David S. F. Portree with John Young, June 13, 1996; Eric Jones, email, August 19, 1996.

April 25
1972 EVA 4
World EVA 26
U.S. EVA 24
Deep Space EVA 2
Duration: 1:24
Spacecraft/mission: Apollo 16
Crew: John Young, Charles Duke, Thomas Mattingly
Spacewalker: Thomas Mattingly
Purpose: Retrieve film from SIM bay of CM *Casper*

During the flight home moondust drifted around *Casper*'s cabin. Some spilled into space when Mattingly stepped out to recover mapping and panoramic camera film from the SIM bay. Mattingly made two leisurely trips along *Casper*'s Service Module. He inspected the spacecraft's exterior and exposed the Microbial Ecological Evaluation Device to space for 10 min. Before returning to the cabin, he opened his visor briefly so he could see the stars, taking care not to look in the direction of the Sun.

Astronautics and Aeronautics 1972, NASA SP 4017, p. 149; *Apollo 16 Preliminary Science Report*, NASA SP 315, 1972, p. 2.10; *A Man on the Moon*, Andrew Chaikin, 1994, p. 492.

April 27   **Apollo 16 splashdown**

December 7   **Apollo 17 launch**

December 11
1972 EVA 5
World EVA 27
U.S. EVA 25
Lunar Surface EVA 13
Duration: 7:12
Spacecraft/mission: Apollo 17
Crew: Eugene Cernan, Harrison Schmitt, Ronald Evans
Moonwalkers: Eugene Cernan, Harrison Schmitt
Purpose: Deploy LRV; deploy ALSEP; geological traverse south to Steno crater in Central Cluster

The first Apollo 17 surface EVA began 4 hr after landing with no television of Cernan's first step, the necessary TV equipment having been omitted to save weight and extend LM *Challenger*'s hover time. Harrison Schmitt, the only geologist to visit the Moon, took a proprietary interest in the Taurus-Littrow site - he stepped onto the surface after Cernan and quipped, "Who's been tracking up my lunar surface?" The astronauts deployed and tested their LRV, then planted a U.S. flag which had hung in Mission Control since Apollo 11. Cernan accidentally knocked off part of one of the LRV's fenders and repaired it with tape. The astronauts set up the ALSEP 185 m (605 ft) northwest of the *Challenger*. Cernan drilled two holes 2.5 m (8.2 ft) deep 11 m (35 ft) apart and inserted two heat flow probes. He also drilled a core sample hole 2.8 m (9.2 ft) deep. Collecting the core required 1 hr - the core device stuck despite the long-handled jack designed to ease removal. Cernan's oxygen consumption climbed rapidly as his pulse hit 145 beats per min. Schmitt extracted the core by throwing his weight on the jack handle, but fell and scattered equipment. In general, the Apollo 17 astronauts treated their EMUs roughly - experience gained on earlier flights left them with little fear of suit damage. Cernan then inserted a cosmic-ray probe into the hole left by the core. Encumbered by his suit, Schmitt at first had trouble picking up rocks, which he admitted was "a very embarrassing thing for a geologist." The damaged LRV fender fell off on the way to the first geological survey station, so the astronauts were showered with dust. Schedule pressure forced deletion of a trek to Emory Crater in favor of a shorter trip to Steno Crater. The astronauts placed 0.45-kg (1-lb) and 0.23-kg (1/2-lb) explosive packages during the traverse. The explosives were set off after Cernan and Schmitt departed and recorded by geophones in the Apollo 17 ALSEP. Cernan drove so Schmitt could do science. The geologist carried a new long-handled scoop which allowed him to sample from the rover seat, saving time. Back at *Challenger*, Cernan reported that his tussle with the core tube bruised his arms and burst blood vessels under his fingernails. Dust tracked into the LM gave Schmitt a mild hayfever attack. However, dust catchers on the floor mitigated some of the dust difficulties experienced by previous crews.

Astronautics and Aeronautics 1972, NASA SP 4017, 1974, p. 416; *To a Rocky Moon*, Donald Wilhelms, 1993, pp. 319-322; *Apollo 17 Preliminary Science Report*, NASA SP 330, 1973, p. 2.4, 6.6-6.7; *Apollo Lunar Surface Journal*, Eric Jones, 1995 (http://www.hq.nasa.gov/office/pao/History/alsj/); *A Man on the Moon*, Andrew Chaikin, 1994, pp. 513; interview, David S. F. Portree with John Young, June 13, 1996.

December 12
1972 EVA 6
World EVA 28
U.S. EVA 26
Lunar Surface EVA 14
Duration: 7:37
Spacecraft/mission: Apollo 17
Crew: Eugene Cernan, Harrison Schmitt, Ronald Evans
Moonwalkers: Eugene Cernan, Harrison Schmitt
Purpose: Geological traverse to South Massif

Apollo 16 CDR John Young developed and radioed to Schmitt and Cernan a procedure for repairing the LRV fender using folded traverse maps and two lamp clamps. After completing the repair, the Apollo 17 explorers set out on their second geological traverse. At survey stops they deployed 0.06-, 0.12-, and 2.8-kg (1/8-, 1/4-, and 6-lb) explosive packages. Cernan abandoned some of the caution shown on Apollo 15 and 16 and drove as fast as he could. The astronauts skirted Camelot and Lara craters, and spent an hour sampling South Massif landslide material at Nansen Crater. They then explored Shorty Crater, which was suspected (at this time) of being a volcanic vent. Schmitt kicked up orange and crimson soil which appeared to confirm this hypothesis, so the astronauts, Mission Control, and Earth-based geologists rapidly adjusted the tight traverse schedule so Schmitt could collect unplanned core and trench samples. Apollo scientific and technical ground-based support was sufficiently refined by this time to permit flexible responses to EVA challenges and opportunities. In their section of the *Apollo 17 Preliminary Science Report* the astronauts comment on this, stating that

> one can conceive of many samples. . . left uncollected at this remarkable locality. However, few of our experiences in the Apollo Program better illustrate the inherent quality of scientific investigation that is possible from the integrated effort of so many in so short a time.

The astronauts saw more orange soil at later stops. The soil turned out later to be ancient volcanic glass blasted to the surface when Shorty was formed about a million years ago, not a sign of recent volcanism as originally hoped. Before returning to *Challenger* Schmitt went back to the

ALSEP site to check the orientation of the Lunar Surface Gravimeter. On this EVA, the longest of the Apollo program, the astronauts drove the LRV for 19 km (11.4 mi).

Astronautics and Aeronautics 1972, NASA SP 4017, 1974, p. 416-417; *To a Rocky Moon*, Donald Wilhelms, 1993, pp. 322-327; *Apollo Lunar Surface Journal*, Eric Jones, 1995 (http://www.hq.nasa.gov/office/pao/History/alsj/); *Apollo 17 Preliminary Science Report*, NASA SP 330, 1973, p. 2.5. 5.16, 6.7; *A Man on the Moon*, Andrew Chaikin, pp. 514.

December 13
1972 EVA 7
World EVA 29
U.S. EVA 27
Lunar Surface EVA 15
Duration: 7:16
Spacecraft/mission: Apollo 17
Crew: Eugene Cernan, Harrison Schmitt, Ronald Evans
Moonwalkers: Eugene Cernan, Harrison Schmitt
Purpose: Traverse to North Massif

The astronauts recovered the cosmic ray detector before starting their final traverse because a small solar flare threatened to flood it with low-energy solar protons. Fortunately the flare was not powerful - neither the thin-walled LM cabin nor their EMUs could protect them from a powerful flare. Cernan and Schmitt traversed to North Massif. They deleted the last geological survey stop to return to the ALSEP to adjust the Lunar Surface Gravimeter, which was still not functioning properly. They then extracted the cosmic-ray neutron probe and set more explosive packages. Finally, they unveiled a plaque on *Challenger*, which read:

> Here man completed his first exploration of the Moon, December 1972 A.D. May the spirit of peace in which we came be reflected in the lives of all mankind.

Schmitt and Cernan expected to be the last humans on the Moon until the late 1980s, so they were eager to keep working, but by EVA closeout Schmitt's hands were so tired from lack of glove mobility during the long EVAs that he could barely move them. *Challenger*'s ascent stage lifted off on December 14 carrying 115 kg (253 lb) of samples and 2120 photos.

Astronautics and Aeronautics 1972, NASA SP 4017, 1974, p. 416-417; *To a Rocky Moon*, Donald Wilhelms, 1993, pp. 327-331; *Apollo Lunar Surface Journal*, Eric Jones, 1995 (http://www.hq.nasa.gov/office/pao/History/alsj/); *Apollo 17 Preliminary Science Report*, NASA SP 330, 1973, p. 2.5, 6.7; *A Man on the Moon*, Andrew Chaikin, pp. 542-543.

December 17
1972 EVA 8
World EVA 30
U.S. EVA 28
Deep Space EVA 3
Duration: 1:07
Spacecraft/mission: Apollo 17
Crew: Eugene Cernan, Harrison Schmitt, Ronald Evans
Spacewalker: Ronald Evans
Purpose: Retrieve film from SIM bay of CM *America*

Ron Evans performed the last deep space EVA to date, making three trips to CM *America*'s SIM bay to retrieve film and floating free for a time on his 7.7-m (23-ft) tether.

Astronautics and Aeronautics 1972, NASA SP 4017, 1974, p. 416-417.

December 19	Apollo 17 splashdown

1973

May 25	Skylab 2 launch

May 25
1973 EVA 1
World EVA 31
U.S. EVA 29
Space Station EVA 1
Duration: 0:40
Spacecraft/mission: Skylab Orbital Workshop (Skylab 1)/Skylab 2
Crew: Charles Conrad, Joseph Kerwin, Paul Weitz
Spacewalker: Paul Weitz
Purpose: Free jammed Skylab solar array

Skylab was damaged about one minute after its May 14 liftoff on a Saturn V. The astronauts reported that solar array wing 2 and most of the meteoroid shield were gone. Solar array wing 1 appeared intact, but a metal strap held it closed. Later analysis determined that a design flaw, an opening at the top of the meteoroid shield, allowed air to enter between the station's skin and the shield during ascent. This created an overpressure which ripped away the shield, which in turn snagged and tore away solar array wing 2. On this date Skylab 2 CDR Charles Conrad, CMP Paul Weitz, and Science Pilot (SPT) Joseph Kerwin rendezvoused with the Skylab Orbital Workshop 6 hr after launch on a Saturn IB rocket. The Skylab astronauts wore a modified Apollo A7LB EMU. When used during EVA from the Skylab Airlock Module, the A7LB featured an Astronaut Life Support Assembly (ALSA) belly-pack instead of a PLSS. The ALSA consisted of a oxygen control unit generally similar in function to the Gemini G4C Ventilation Control Module; a 18.3-m (60-ft) insulated umbilical; and a leg-mounted package holding a 30-min emergency oxygen supply in two tanks. Oxygen and cooling water were pumped from the Airlock Module through the umbilical to the ALSA control unit, which distributed them to hose connectors on the front of the suit. For this EVA, however, the astronauts used shorter umbilicals to link them to the CM life support system. The astronauts moved the Skylab 2 CM close to the jammed array. Weitz then stood with his upper body through the hatch and assembled a 4.5-m (15-ft) pole with a shepherd's hook on the end from three 1.5-m (5-ft) sections handed to him by Kerwin. He hooked and pulled on the array while Kerwin gripped his legs. Conrad had to hold the CM steady because Weitz's efforts pulled it toward the workshop. Weitz replaced the hook with a universal prying tool when the strap did not budge, but to no avail. Their efforts thwarted, the astronauts docked with Skylab and closed out a 22-hr day. Conrad was blunt about the likelihood of freeing array wing 1 - he told Mission Control that "we ain't gonna do it with the tools we got." Once inside the station, the Skylab 2 crew deployed a solar shield parasol through a small scientific airlock. They commenced their research program on May 29, but the four Apollo Telescope Mount (ATM) "windmill" arrays proved insufficient to maintain the station.

Living and Working in Space: A History of Skylab (NASA SP-4208), W. David Compton and Charles D. Benson, 1983, pp. 269-271; "Skylab EVA," Robert Kain, Crew Training and Procedures Division, NASA

JSC, no date; *Skylab Experience Bulletin No. 27: Personnel and Equipment Restraint and Mobility Aids: EVA.* JSC 09561, NASA JSC, May 1975.

June 7
1973 EVA 2
World EVA 32
U.S. EVA 30
Space Station EVA 2
Duration: 3:25
Spacecraft/mission: Skylab Orbital Workshop (Skylab 1)/Skylab 2
Crew: Charles Conrad, Joseph Kerwin, Paul Weitz
Spacewalkers: Charles Conrad, Joseph Kerwin
Purpose: Free jammed Skylab solar array

A team led by Russell Schweickart developed an EVA solar array repair procedure, which NASA management approved on June 4. Mission controllers sent EVA instructions to Skylab's tele-printer and the astronauts fabricated tools from onboard materials. They screwed together six 1.5-m (5-ft) rods attached a cable cutter at the other, then tied 6 m (20 ft) of rope from the SEVA Sail backup solar shield to the cutter pull rope. This permitted the EVA astronaut to operate the cutter from 9 m (30 ft) away - the distance from the edge of area around Skylab's airlock hatch (the EVA Bay) to the strap holding shut array wing 1. An EVA waist tether was hooked to the cable cutter assembly to attach it to the base of Skylab's discone antenna. The cable cutter assembly also served as a handrail for translation to the solar array wing. Engineers believed that a hydrau-lic damper for slowing normal wing deployment had frozen, so Schweickart's team devised the Beam Erection Tether (BET) to force it open. The BET was a 9.8-m (32-ft) piece of SEVA sail rope tied to the middle of a 1.8-m (6-ft) rope. Two small hooks from waist tethers were tied to the ends of the 1.8-m rope, and a large hook designed originally to secure the SEVA sail to Skylab's exterior was tied onto the opposite end of the 9.8-m rope. The two small hooks were attached to holes in the wing array, while the large hook was attached near the discone antenna. Because footholds were scarce, the astronauts could not deploy the array by pulling on one end of the BET. Instead, they would stand with the middle of the BET over one shoulder to hold them against Skylab's hull and break the damper by straining upwards and pulling. The BET would thus served both to open the array and restrain the astronauts. On June 6 Kerwin and Conrad rehearsed the planned EVA inside Skylab, and on this date depressurized the Skylab Airlock Module, which was made cramped by their burden of tools. Conrad left the airlock through its surplus Gemini hatch and stepped into the Pressure Garment Assembly foot restraint at the Fixed Airlock Shroud work station. Kerwin passed him six 1.5-m (5-ft) poles, helped him assemble the cable cutter assembly, then moved to the discone antenna using the ATM girders and other projections in the EVA Bay as mobility aids. Conrad handed him the cable cutter assembly, then moved to the discone antenna carrying the BET. The plan called for Kerwin to hook the cable cutter assembly on the strap holding wing 1 closed. Conrad would then crawl down the assembly to wing 1 and attach the BET. However, Kerwin had difficulties finding a firm foothold near the discone because Skylab unexpectedly differed from the mockup in the tank in Huntsville. He was forced to hold on with one hand while attempting to position the pole with the other. After a frustrating half hour, Kerwin shortened his 1.8-m (6-ft) tether by doubling it. This held him more firmly against Skylab and allowed him partial use of his other hand. He finally succeeded in hooking the aluminum strap. Conrad attached the BET large hook to the discone antenna, then climbed along the cable cutter assembly pole. He attached one of the two BET small hooks to bolt holes on wing 1. Again the flight Skylab differed from the ground mockup; the second small hook would not fit. Kerwin tightened the BET at the discone end using a cleat, then cut the strap holding the array closed. Conrad placed the BET over his shoulder, put his feet against the workshop's hull, and strained against the BET to pull open the array. Kerwin joined him. Finally the hydraulic damper holding the array closed gave way. As Conrad later described it: "I was facing away from it, heaving with all my might, and Joe was also heaving with all his might when it let go and both of us took off. By the time we got settled down. . . those panels were out as far as they were going to go." Needles on electricity meters on the ground and inside Skylab jumped, signaling success. The astronauts serviced the ATM before going inside, changing out film in a malfunctioning camera and pinning open a balky solar telescope aperture door. The astronauts had difficulty restowing the life support umbilicals in their spherical stowage containers. The primary EVA heat exchanger module suffered minor clogging during the EVA, leading engineers to design a new module to serve as a backup. The module reached the station in July with the second Skylab crew.

Living and Working in Space: A History of Skylab (NASA SP-4208), W. David Compton and Charles D. Benson, 1983, pp. 272-276; "Skylab EVA," Robert Kain, Crew Training and Procedures Division, NASA JSC, no date; *Skylab Experience Bulletin No. 27: Personnel and Equipment Restraint and Mobility Aids: EVA*, JSC 09561, NASA JSC, May 1975; "The Skylab Missions," *Marshall Star*, May 11, 1988; "Record Payload for Next Skylab," *Aviation Week & Space Technology*, July 2, 1973, p. 16; *Personal Logs*, Joseph McMann.

June 19
1973 EVA 3
World EVA 33
U.S. EVA 31
Space Station EVA 3
Duration: 1:36
Spacecraft/mission: Skylab Orbital Workshop (Skylab 1)/Skylab 2
Crew: Charles Conrad, Joseph Kerwin, Paul Weitz
Spacewalkers: Charles Conrad, Paul Weitz
Purpose: Replace ATM film; repair circuit breaker module

Before Skylab was launched, the Skylab 2 flight plan called for Conrad and Kerwin to perform one 2 hr, 30 min EVA on mission day 26. As it turned out, the EVA was actually the third of the flight, and was carried out by Conrad and Weitz. The first part of the EVA was very similar to the EVA as planned pre-flight. The astronauts removed film from the ATM solar telescopes for return to Earth and replaced the film. This required a fraction of the time planned. Four of Skylab's five EVA work stations were positioned to restrain the astronaut during film changeout. The Fixed Airlock Shroud (FAS) station, the main EVA work station, was located next to the external airlock hatch. The FAS station was the "base camp" for ascending the ATM. The astronauts moved between the work stations via the Deployment Assembly route, or "EVA Trail." The route consisted of single and dual handrails, the latter resembling ladders without rungs. According to the Skylab astronauts, the single handrails worked well, while translation using the dual rails was as easy as "driving on the freeway." All handrails were painted blue for visibility and provided with "road signs" - alphanumeric designators. The blue faded rapidly in the strong sunlight of space, however, and the designator labels proved difficult to see. ATM film cassettes were moved in a device called a film tree. The primary method of moving the trees was by three extendible booms located in the EVA Bay within reach of FAS. Controls for the motorized booms were located next to the EVA hatch. The booms could be manually operated if necessary, and "clothes-lines," pulley-type devices, provided a backup film transport method. Their film changeout tasks completed, Conrad and Weitz removed space exposure samples launched on the workshop's exterior to accompany them back to Earth. Weitz and Conrad then moved on to tasks added after the successful wing 1 deployment on EVA 2. They used a brush to clean the White Light Corono-graph occulting disk, which was producing glare. Conrad then moved to Circuit Breaker Relay

Module 15. Acting on instructions from the ground, he hit it with a hammer to free a stuck relay. This low-tech solution succeeded and soon the module was feeding electricity into the Skylab power system again. The EVA brought the total for Skylab 2 to more than 5 hr, twice what was originally planned.

Living and Working in Space: A History of Skylab (NASA SP-4208), W. David Compton and Charles D. Benson, 1983, p. 294; "Skylab EVA," Robert Kain, Crew Training and Procedures Division, NASA JSC, no date; *Skylab Experience Bulletin No. 27: Personnel and Equipment Restraint and Mobility Aids: EVA*, JSC 09561, NASA JSC, May 1975.

| June 22 | **Skylab 2 splashdown** |
| July 28 | **Skylab 3 launch** |

August 6
1973 EVA 4
World EVA 34
U.S. EVA 32
Space Station EVA 4
Duration: 6:31
Spacecraft/mission: Skylab Orbital Workshop (Skylab 1)/Skylab 3
Crew: Alan Bean, Owen Garriott, Jack Lousma
Spacewalks: Owen Garriott, Jack Lousma
Purpose: Install new sunshade; replace ATM film

This EVA was scheduled before launch to occur on mission day 4, but crew illness (space motion sickness) pushed it back to mission day 10. The main order of business was to install the Twin Pole Sunshade over the parasol installed by the Skylab 2 astronauts because testing on the ground showed that the parasol's nylon fabric could deteriorate from exposure to solar ultraviolet radiation. SPT Garriott assembled two poles, each made up of 11 1.5-m (5-ft) sections, and passed them to CMP Lousma, who was positioned in the portable foot restraint attached to an ATM handrail. Lousma attached the poles to a base plate he installed on a hand rail, unfurled the sunshade fabric, and attached a reefing line to make the shade lie flat. He swiveled the completed shade to cover the station, then returned to the airlock to get equipment for the next phase of the EVA. Lousma ascended the ATM again, installed film, then inspected thruster quads A and B on the Skylab 3 CM from his position on the ATM. The quads were leaking, but Lousma saw no obvious signs of leakage and they later stopped, so Skylab 3 could run its scheduled 56-day duration. Lousma removed a telescope aperture door ramp to keep the door from sticking, which required removal of two bolts not designed for EVA, then deployed the Micrometeoroid Particle Collection experiment. The experiment was originally intended for deployment from the science airlock blocked by the parasol, but was redesigned for EVA deployment and launched with the Skylab 3 crew.

"Skylab EVA," Robert Kain, Crew Training and Procedures Division, NASA JSC, no date; *Skylab: A Chronology*, NASA SP 4011, Roland Newkirk, *et al*, 1977, pp. 324-325; *Skylab Experience Bulletin No. 27: Personnel and Equipment Restraint and Mobility Aids: EVA*, JSC 09561, NASA JSC, May 1975.

August 24
1973 EVA 5
World EVA 35
U.S. EVA 33
Space Station EVA 5

Duration: 4:31
Spacecraft/mission: Skylab Orbital Workshop (Skylab 1)/Skylab 3
Crew: Alan Bean, Owen Garriott, Jack Lousma
Spacewalkers: Owen Garriott, Jack Lousma
Purpose: Change out all ATM film; install rate gyro package cable

Before Skylab reached orbit, this EVA to change out all ATM film was scheduled to last 2 hr, 45 min on day 29 of Skylab 3. The EVA actually occurred on mission day 28. In addition to their film changeout task, Garriott and Lousma installed a 7.3-m (24-ft) cable for a new rate gyro package they installed within the station's pressurized volume. The astronauts also attached a clipboard with two parasol material samples to a handrail, and removed two more ramps from faulty ATM aperture doors.

"Skylab EVA," Robert Kain, Crew Training and Procedures Division, NASA JSC, no date; *Skylab Experience Bulletin No. 27: Personnel and Equipment Restraint and Mobility Aids: EVA*, JSC 09561, NASA JSC, May 1975.

September 22
1973 EVA 6
World EVA 36
U.S. EVA 34
Space Station EVA 6
Duration: 2:41
Spacecraft/mission: Skylab Orbital Workshop (Skylab 1)/Skylab 3
Crew: Alan Bean, Owen Garriott, Jack Lousma
Spacewalkers: Alan Bean, Owen Garriott
Purpose: Remove all ATM film, partially replace; remove space exposure samples and collectors

Skylab 3 astronauts Bean and Garriott removed all ATM film for return to Earth, performed partial ATM film installation, and retrieved exposed collectors and samples, including one parasol material sample from the clipboard. The Airlock Module suit cooling system was inoperative because of leaks, so no water flowed through the umbilicals to the astronauts' suits. Air cooling proved adequate for the undemanding tasks at hand, Garriott reported becoming slightly warm, while Bean's hands were warm throughout the EVA.

"Skylab EVA," Robert Kain, Crew Training and Procedures Division, NASA JSC, no date; *Skylab Experience Bulletin No. 27: Personnel and Equipment Restraint and Mobility Aids: EVA*, JSC 09561, NASA JSC, May 1975; *Personal Logs*, Joseph McMann.

September 25	Skylab 3 splashdown
September 27-29	Soyuz 12
November 16	Skylab 4 launch

November 22
1973 EVA 7
World EVA 37
U.S. EVA 35
Space Station EVA 7
Duration: 6:33

Spacecraft/mission: Skylab Orbital Workshop (Skylab 1)/Skylab 4
Crew: Gerald Carr, Edward Gibson, William Pogue
Spacewalkers: William Pogue, Edward Gibson
Purpose: Take Earth atmosphere photos; install experiments; repair Microwave Radiometer/Scatterometer/Altimeter antenna

Before Skylab was launched, Skylab 4 SPT Gibson and CMP Pogue were to retrieve a meteoroid collector and partially install ATM film during their first EVA, which was to last 2 hr on mission day 4. The EVA actually occurred on mission day 7 and lasted three times as long. The crew refilled the Airlock Module suit cooling system with water prior to the EVA. In addition to carrying out the originally scheduled tasks, Pogue and Gibson placed the Coronagraph Contamination Measurements experiment on an ATM truss and attempted to photograph Earth's atmosphere using a camera originally intended for deployment from the science airlock blocked by the parasol solar shield. The camera failed after 5 of 40 planned exposures. They also attached the Trans-Uranic Cosmic Ray Experiment detector to the clipboard; pinned open a malfunctioning aperture door; installed space exposure samples; and repaired the Microwave Radiometer/Scatterometer/Altimeter antenna, which was on the Earth-facing side of station where no EVA handrails or foot restraints existed. The astronauts had difficulty keeping their umbilicals separated.

"Skylab EVA," Robert Kain, Crew Training and Procedures Divisions, NASA JSC, no date; *Skylab Experience Bulletin No. 27: Personnel and Equipment Restraint and Mobility Aids: EVA*, JSC 09561, NASA JSC, May 1975; *Personal Logs*, Joseph McMann.

December 18-26   **Soyuz 13**

December 25
1973 EVA 8
World EVA 38
U.S. EVA 36
Space Station EVA 8
Duration: 7:01
Spacecraft/mission: Skylab Orbital Workshop (Skylab 1)/Skylab 4
Crew: Gerald Carr, Edward Gibson, William Pogue
Spacewalkers: Gerald Carr, William Pogue
Purpose: Perform partial ATM film replacement; photograph Comet Kohoutek; repair ATM telescope

During this, the longest Skylab EVA, Pogue and CDR Carr attached the X-ray/ultraviolet (UV) Solar Photography experiment to the ATM truss. The experiment was originally intended for deployment from the science airlock blocked by the parasol. They also took 40 pictures of Comet Kohoutek; partially replaced ATM film; retrieved space exposure samples; and pinned open another malfunctioning aperture door. The astronauts then returned to the Airlock Module to stowed equipment while IV crewman Edward Gibson maneuvered Skylab to the proper attitude for far UV comet photography. The astronauts attached the Far UV camera to an ATM truss, took three sequences of 10 photos each, then returned the instrument to the airlock. Finally they repaired a telescope filter wheel, which involved fine work made challenging by their stiff space suit gloves. The astronauts used a dental mirror and penlight to look into the aperture, then carefully positioned the filter wheel with a screwdriver. Leaking cooling water, colored yellow by chromate corrosion inhibitor, formed ice on the front of Carr's belly-mounted pressure control unit. The leak lacked sufficient volume to deplete the cooling water supply. Yellow ice flaked off as Carr moved.

"Skylab EVA," Robert Kain, Crew Training and Procedures Division, NASA JSC, no date; *Skylab Experience Bulletin No. 27: Personnel and Equipment Restraint and Mobility Aids: EVA*, JSC 09561, NASA Johnson Space Center, May 1975; *Personal Logs*, Joseph McMann.

December 29
1973 EVA 9
World EVA 39
U.S. EVA 37
Space Station EVA 9
Duration: 3:29
Spacecraft/mission: Skylab Orbital Workshop (Skylab 1)/Skylab 4
Crew: Gerald Carr, Edward Gibson, William Pogue
Spacewalkers: Gerald Carr, Edward Gibson
Purpose: Retrieve space exposure samples; photograph Comet Kohoutek

On mission day 44 Carr and Gibson collected a piece of the Airlock Module meteoroid cover for analysis. They then repeated their Comet Kohoutek observations. During the EVA, ice formed on the front of Carr's suit because of a cooling water leak.

"Skylab EVA," Robert Kain, Crew Training and Procedures Division, NASA JSC, no date; *Skylab Experience Bulletin No. 27: Personnel and Equipment Restraint and Mobility Aids: EVA*, JSC 09561, NASA JSC, May 1975.

1974

February 3
1974 EVA 1
World EVA 40
U.S. EVA 38
Space Station EVA 10
Duration: 5:19
Spacecraft/mission: Skylab Orbital Workshop (Skylab 1)/Skylab 4
Crew: Gerald Carr, Edward Gibson, William Pogue
Spacewalkers: Gerald Carr, Edward Gibson
Purpose: Collect space exposure samples and collectors; remove all Skylab film for return to Earth

Originally the Skylab 4 crew was to have undertaken only two EVAs, the last occurring on mission day 54, a few weeks before their return to Earth. The only task slated for the EVA was retrieval of all ATM film. The actual last Skylab 4 EVA, the fourth of the mission, took place on mission day 80, days before the crew's return to Earth, and included 16 tasks. Gibson and Carr used the backup "clothesline" film transfer device to move film, collectors, and a camera back and forth between the ATM and the Airlock Module. Carr demonstrated hand over hand movement along a tether. The astronauts completed the Earth atmosphere photography begun on their first EVA by snapping photographs between their ATM film removal tasks. The astronauts then mounted the Micrometeoroid Particle Collection experiment on the ATM. NASA hoped that the experiment could be collected by Space Shuttle astronauts during a Skylab visit in the early 1980s. Gibson's suit cooling system sprang a leak, so he switched to minimum cooling and continued work. At the end of the EVA he reported that he was "tired and hungry."

"Skylab EVA," Robert Kain, Crew Training and Procedures Division, NASA JSC, no date; *Skylab Experience Bulletin No. 27: Personnel and Equipment Restraint and Mobility Aids: EVA*, JSC 09561, NASA JSC, May 1975.

February 8	**Skylab 4 splashdown**
July 3-19	**Salyut 3/Soyuz 14 Expedition 1**
August 26-28	**Salyut 3/Soyuz 15**
December 2-8	**Soyuz 16**

1975

January 10-February 9	**Salyut 4/Soyuz 17 Expedition 1**
May 24-July 28	**Salyut 4/Soyuz 18 Expedition 2**
July 15-24	**Apollo "18" (Apollo-Soyuz Test Project)**
July 15-21	**Soyuz 19 (Apollo-Soyuz Test Project)**

1976

July 6-August 24	**Salyut 5/Soyuz 21 Expedition 1**
September 15-23	**Soyuz 22**
October 14-16	**Salyut 5/Soyuz 23**

1977

February 7-25	**Salyut 5/Soyuz 24 Expedition 2**
October 9-11	**Salyut 6/Soyuz 25**
December 10	**Salyut 6/Soyuz 26 Principal Expedition (PE) 1 launch**

December 20
1977 EVA 1
World EVA 41
Russian EVA 3
Space Station EVA 11
Duration: 1:28
Spacecraft/mission: Salyut 6 PE-1
Crew: Yuri Romanenko, Georgi Grechko
Spacewalkers: Yuri Romanenko, Georgi Grechko
Purpose: Test Orlan-D space suit in depressurized airlock; inspect Salyut 6 front port

This SEVA, the first Soviet EVA since 1969, was originally planned as a test of the Orlan-D suit in the depressurized transfer compartment at the front of Salyut 6. The Orlan-D was based on the Orlan lunar EVA suit, the most distinctive features of which were a hard torso, adjustable soft limbs, and simple self-donning via a hatch in the back. The hatch cover contained life support equipment, removing the need for external hoses. The Orlan lunar suit was designed to be used by one cosmonaut on a single mission. Orlan-D, as the redesigned suit was called, was to remain aboard a station for up to 2 years and be used by several cosmonauts. The suit operated at 40 kpascal (5.8 psi), permitting a pure oxygen prebreathe period of only 30 minutes. For Salyut 6, EVA duration was limited to about 3 hr. A waist tether with a "snap lock" tether hook for attaching to handrails outside the station was considered integral to the suit. The Orlan-D relied for electrical power and voice communications on an umbilical plugged into a socket in the space station transfer compartment. Because Soyuz 25 could not dock at Salyut 6's front port, Flight Engineer Grechko was given the additional task of inspecting and recertifying the port for future Soyuz Ferry dockings. Grechko opened the front docking port and pulled himself halfway out so that he could inspect and manipulate the outer surfaces of the docking mechanism using special tools. He found everything to be in perfect working order. For years Soviet spaceflight observers believed that mission Commander Romanenko, in his eagerness to look out the open hatch, nearly drifted free of the station, and that only quick action by Grechko prevented him from being lost in space. Grechko now denies categorically that his commander was ever in danger, and adds ruefully that "Yuri was very angry about the story." Romanenko says that the story had its start in a "bad joke" Grechko told which was misunderstood, and adds that, even though his short safety tether was not secured, his electricity/communications umbilical firmly fastened him to Salyut 6. Depressurized time was 88 min, but SEVA time was only 20 min.

Red Star in Orbit, James E. Oberg, Random House, 1981, pp. 165-167; "Meeting Cosmonaut Georgi Grechko," George Spiteri and Tony Bird, *Spaceflight*, July 1994, p.245; "Romanenko - Living and Working in Space," *Spaceflight*, September 1989, p. 294; "Above the Planet: Salyut EVA Operations," Neville Kidger, *Spaceflight*, February 1989, pp. 48-49; "Space Suits: Ten Periods of Extravehicular Activity from the Salyut 7 Space Station," G.I. Severin, *et al*, 35th IAF Congress, October 7-13, 1984, pp. 2-3; *Nauchnyy Orbital'nyy Kompleks*, Konstantin Feoktistov, *Novoye v Zhizhni, Nauke, Tekhnike, Seriya Kosmonavtika, Astronomiya*, No. 3, 1980, pp. 1-63 (translated in *USSR Report: Space*, JPRS L/9145, June 17, 1980, p. 10); "211 Days in Orbit," V. Gorkoy and N. Konkov, *Aviatsiya i Kosmonavtika*, August 1983, pp. 40-41 (translated in *USSR Report: Space*, JPRS-USP-84-006-L, July 20, 1984, pp. 61-62); "The Experience in Operation and Improving the Orlan-type Space Suit," I. P. Abramov, *Acta Astronautica*, Vol. 36, No. 1, July 1995, pp. 1-12; *Pressure Suits and Systems for Working in Open Space*, I. P. Abramov, G. I. Severin, *et al*, Machinostroyeniye Publishing House, Moscow, 1984 (translated by U.S. Air Force Systems Command, Foreign Technology Division, March 25, 1987).

1978

January 10	Salyut 6/Soyuz 27 Visiting Expedition (VE) 1 launch
January 16	Salyut 6/Soyuz 26 VE 1 landing
March 2-10	Salyut 6/Soyuz 28 VE 2
March 16	Salyut 6/Soyuz 27 PE 1 landing
June 15	Salyut 6/Soyuz 29 PE 2 launch
June 27-July 5	Salyut 6/Soyuz 30 VE 3

July 29
1978 EVA 1
World EVA 42
Russian EVA 4
Space Station EVA 12
Duration: 2:05
Spacecraft/mission: Salyut 6 PE-2
Crew: Vladimir Kovalyonok, Alexandr Ivanchenkov
Spacewalkers: Vladimir Kovalyonok, Alexandr Ivanchenkov
Purpose: Remove space exposure cassettes and detectors from hull

Forty-five days into their 140-day stay on Salyut 6, Kovalyonok and Ivanchenkov don the Orlan-D suits first worn by Grechko and Romanenko. They then depressurized the transfer compartment and open the hatch in its side, starting the first full-emergence EVA of the Soviet space program since 1969. The transfer compartment was located at the front of the Salyut 6 station. The 2-m-dia (6.56-ft-dia) cylindrical compartment included a round hatch leading out onto the station's port side. An air-tight hatch at the front separated the compartment from the docked Soyuz, while another at the rear sealed off the 4.15-m-dia (13.6-ft-dia) cylindrical work compartment. The transfer compartment contained valves to spill its air into space; valves to refill it with air from the work compartment; control and display panels; connectors for umbilicals providing electricity and communications to the suits; anchoring points for restraints and tethers; and storage compartments for tethers, foot restraints, two Orlan-D suits, and other EVA equipment. Handrails on the station's exterior converged at the airlock hatch. Flight Engineer Ivanchenkov positioned himself on the Yakor ("anchor") foot restraint near the airlock hatch, while Kovalyonok floated with his feet in the transfer compartment. The cosmonauts rested during the 35-min orbital night, and were treated to the sight of a brilliant meteor burning up below them in Earth's atmosphere. Ivanchenkov removed three space exposure cassettes launched on Salyut 6's exterior and handed them to Kovalyonok for stowage in the transfer compartment. He then replaced meteoroid dust collectors and installed radiation sensors. Ivanchenkov photographed the Black Sea, Kazakhstan, and China. The *Tsentr Upravleniya Polyotami* ("Flight Control Center") (TsUP) in Kaliningrad, outside Moscow, ordered the spacewalkers to return inside just before Salyut 6 passed from radio range. Kovalyonok decided while out of radio contact that they could stretch the EVA 20 min longer to enjoy the view of Australia's Great Barrier Reef and New Zealand. After the EVA, the cosmonauts used air from the Progress 2 logistic resupply ship's tanks to replace that vented when they depressurized the transfer compartment. Using Progress freighters to make up EVA air loss becomes standard practice.

"Space Suits: Ten Periods of Extravehicular Activity from the Salyut 7 Space Station," G.I. Severin, *et al*, 35th IAF Congress, October 7-13, 1984, pp. 2-3; *Nauchnyy Orbital'nyy Kompleks*, Konstantin Feoktistov, *Novoye v Zhizhni, Nauke, Tekhnike, Seriya Kosmonavtika, Astronomiya*, No. 3, 1980, pp. 1-63 (translated in *USSR Report: Space*, JPRS L/9145, June 17, 1980, p. 10); "211 Days in Orbit," V. Gorkoy and N. Konkov, *Aviatsiya i Kosmonavtika*, August 1983, pp. 40-41 (translated in *USSR Report: Space*, JPRS-USP-84-006-L, July 20, 1984, pp. 61-62); "The Experience in Operation and Improving the Orlan-type Space Suit," I. P. Abramov, *Acta Astronautica*, Vol. 36, No. 1, July 1995, pp. 1-12; *Pressure Suits and Systems for Working in Open Space*, I. P. Abramov, G. I. Severin, *et al*, Machinostroyeniye Publishing House, Moscow, 1984 (translated by U.S. Air Force Systems Command, Foreign Technology Division, March 25, 1987); *Red Star in Orbit*, James E. Oberg, Random House, 1981, pp. 202-204; "Above the Planet: Salyut EVA Operations," Neville Kidger, *Spaceflight*, February 1989, pp. 48-49.

August 26-September 3   **Salyut 6/Soyuz 31 VE-4**

November 2   **Salyut 6/Soyuz 31 PE-2 landing**

1979

August 15
1979 EVA 1
World EVA 43
Russian EVA 5
Space Station EVA 13
Duration: 1:23
Spacecraft/mission: Salyut 6 PE-3
Crew: Valeri Ryumin, Vladimir Lyakhov
Spacewalkers: Vladimir Lyakhov, Valeri Ryumin
Purpose: Discard KRT-10 radio telescope dish

On day 172 of Salyut 6 PE-3, Lyakhov and Ryumin transferred their experiment results, film, specimen cassettes, and personal items to the Soyuz 34 spacecraft, in case they were unable get back inside the Salyut 6 work compartment after this contingency EVA to remove the KRT-10 antenna from Salyut 6's aft docking port. During EVA preparations the backup fan in Ryumin's suit failed because its control unit was exposed to high humidity. Insulation on the control unit was subsequently improved in later suits. Soviet surgeons reported later that Ryumin and Lyakhov maintained a heart rate very similar to their preflight rate (about 60 beats/min) throughout the flight, but that this changed during the EVA. Ryumin's heart rate reached 130 beats/min while he waited in the airlock. As flight engineer, he clambered out the EVA hatch first, about 10 min before Salyut 6 slipped into night. By this time the cosmonauts were out of range of ground stations. Three hr were allotted for the EVA, and communication was expected to be restored halfway through. With difficulty, Ryumin opened a handrail recessed into the station's hull, then gripped it with one hand for 30 min until orbital sunrise. Lyakhov remained inside the depressurized transfer compartment. As Ryumin moved aft along the station, Lyakhov took his place at the handrail and paid out the umbilical connected to Ryumin's Orlan-D suit. Ryumin had to use a "nipper" (wirecutters designed for IV use) to cut through four 1-mm steel cables and free the antenna, which flops loosely at the rear of the station. Each time he cut at a cable, the 10-m-dia (33-ft-dia) antenna pitched toward him, threatening to cut his suit (as it had cut thermal insulation blankets on the station's aft bulkhead) or smash him. Lyakhov positioned himself so he could warn Ryumin of the dish's movements. Ryumin's heart rate reached 146 beats/min while he worked. Finally, the last cable was cut and Ryumin pushed the KRT-10 antenna away with a "forked stick." When communication was restored with the ground the cosmonauts reported that they had completed their task, and at first were not believed. While flight controllers and official spectators in the TsUP applauded their triumph, Ryumin and Lyakhov returned to the exterior of the transfer compartment, where Ryumin wiped a porthole with a cloth to collect samples of obscuring "space dust." The cosmonauts also collected space exposure experiment cassettes mounted just outside the EVA hatch. After the EVA Ryumin found a small puncture in his suit's primary bladder, possibly caused by a sharp wire on the KRT-10 antenna.

A Year Away from Earth: A Cosmonaut's Diary, Valeri Ryumin, Molodoya Gvardiya, 1987 (translated in *JPRS Report, Science & Technology, U.S.S.R.: Space*, February 12, 1990); "Above the Planet: Salyut EVA Operations," Neville Kidger, *Spaceflight*, March 1989, pp. 102-103; "Salyut 6 Mission Report: Part 2," Neville Kidger, *Spaceflight*, p. 112; *Vestnik Akademii Nauk SSSR*, O.G. Gazenko and A.D. Yegorov, No. 9, September 1980, pp. 49-59 (translated in *USSR Report: Space*, JPRS L/9526, February 5, 1981, pp. 1-11); "The Experience in Operation and Improving the Orlan-type Space Suits," I. P. Abramov, *Acta Astronautica*, Vol. 36, No. 1, July 1995, pp. 1-12; *Mir Hardware Heritage* (NASA RP 1357), David S. F. Portree, March 1995.

August 19	Salyut 6/Soyuz 34 PE-3 landing

1980

April 9	Salyut 6/Soyuz 35 PE-4 launch
May 26	Salyut 6/Soyuz 36 VE-6 launch
June 3	Salyut 6/Soyuz 35 VE-6 landing
June 5-9	Salyut 6/Soyuz-T 1 VE-7
July 23	Salyut 6/Soyuz 37 VE-8 launch
July 31	Salyut 6/Soyuz 36 VE-8 landing
September 18-26	Salyut 6/Soyuz 38 VE-9
October 11	Salyut 6/Soyuz 39 PE-4 landing
November 27-December 10	Salyut 6/Soyuz-T 3 PE-5

1981

March 12-May 26	Salyut 6/Soyuz-T 4 PE-6
March 22-30	Salyut 6/Soyuz 39 VE-10
April 12-14	STS-1/Columbia
May 14-22	Salyut 6/Soyuz 40 VE-11
November 12-14	STS-2/Columbia

1982

March 22-30	STS-3/Columbia
May 13	Salyut 7/Soyuz-T 5 PE-1 launch
June 24-July 2	Salyut 7/Soyuz-T 6 VE-1

July 30
1982 EVA 1
World EVA 44
Russian EVA 6
Space Station EVA 14
Duration: 2:33
Spacecraft/mission: Salyut 7 PE-1
Crew: Valentin Lebedev, Anatoli Berezevoi
Spacewalkers: Valentin Lebedev, Anatoli Berezevoi
Purpose: Replace space exposure cassettes; test assembly techniques; test upgraded Orlan-D space suit

This was the first Soviet EVA since August 1979. Lebedev and Berezevoi, the first Salyut 7 crew, wore suits sufficiently similar to the Salyut 6 Orlan-D that they received no new designation. The Salyut 7 Orlan-D did, however, incorporated improvements based on Salyut 6 EVA experience. For example, external connectors were added to supply the cosmonauts with air and cooling water through an umbilical connected to the Salyut 7 life support system while they were in the transfer compartment airlock. This permitted them to avoid using their finite suit supplies until they were ready to venture outside. In addition, the suit controls were "more conveniently located on the chest," there was an improved cooling system, and EVA duration was extended to 5 hr. July 27 was a refresher training day, when Flight Engineer Lebedev and Commander Berezevoi donned and checked out their suits. Lebedev had some trouble getting through the rear suit hatch - the struggle left him breathless - and was surprised to discover that weightlessness offered no help. The EVA took place on day 78 of the expedition. In his diary, Lebedev described EVA preparations:

> After we got up at 10 p.m., we ate breakfast and did medical tests. My blood pressure was 106/86 and my pulse was 100. . . the result of a sleepless night. Then we re-oriented the station, deactivated the gyroscopes. . . and switched it over to the Kaskad [automatic stabilization control system]. After that our station was in a fixed position pointing toward the stars. We put on our undergarment accessories and prepared our transport vehicle [Soyuz-T 5] for an emergency departure. . . We deactivated the station, closed the hatch between the station and the transport vehicle as well as the hatch between the [transfer] and working compartments. We put on our suits. Ground Control told us to be completely dressed by 3:50 a.m. . .

Lebedev and Berezevoi dumped the transfer compartment's air into space, then opened the round hatch in the compartment's side. Escaping residual air carried dust and debris, including a pencil, through the open hatch into space. Berezevoi said later that the light outside was like "being on a street on a bright sunny day with the ground covered in pure white snow." Berezevoi spent most of the EVA standing in the hatch passing equipment to Lebedev, who positioned himself near the hatch in the Yakor foot restraint. Lebedev's activities were aimed at preparing for Salyut 7 solar array augmentation EVAs and other space assembly tasks. The Pamyat experiment tested "thermomechanical joining of pipeline sections in outer space," while Istok tested "threaded connectors" made of different materials. Lebedev also tested an experimental wrench, which worked well, but his hand went numb because his wrist pressed against his suit wrist ring during tool use. Lebedev complained that the improved Orlan-D cooling system made his feet cold. He collected and replaced 20 space exposure cassettes containing gasket rubber, insulating coatings, glass for ports and lenses, and other materials. He also replaced micrometeoroid detectors.

Lebedev then described the EVA for Soviet TV viewers as he stood on Salyut 7's hull while Berezevoi televised the station's exterior and the Earth below.

"Chronology of Salyut 7 Flight: July 30," *Pravda*, July 31, 1982, p. 1 (translated in *USSR Report: Space*, No. 18, JPRS 82169, November 4, 1982, pp. 3-4); "Above the Planet: Salyut EVA Operations," Neville Kidger, *Spaceflight*, March 1989, pp. 104; "The Hatch is Thrown Open to the Universe," *Pravda*, A. Pokrovski, July 31, 1982, pp. 1, 3 (translated in *USSR Report: Space*, No. 20, JPRS 82970, February 28, 1983, pp. 36-37); "Salyut 7 Cosmonaut Press Conference," Moscow Domestic Service, January 6, 1983 (translated in *USSR Report: Space*, No. 21, JPRS 83430, May 9, 1983, pp. 9-10); *Diary of a Cosmonaut*, Valentin Lebedev, 1990, pp. 142-144, 149-157; "Anatoli Nikolayevich Berezevoi Memoirs: 211 Days in Orbit," *Aviatsiya i Kosmonavtika*, No. 8, August 1983 (translated in *USSR Report: Space*, JPRS-USP-84-006-L, July 20, 1984, pp. 60-62).

August 19	**Salyut 7/Soyuz-T 7 VE-2 launch**
August 27	**Salyut 7/Soyuz-T 5 VE-2 landing**
November 11-16	**STS-5/Columbia**
December 10	**Salyut 7/Soyuz-T 7 PE-1 landing**

1983

April 4   **STS-6/Challenger launch**

April 7
1983 EVA 1
World EVA 45
U.S. EVA 39
Shuttle EVA 1
Duration: 4:10
Spacecraft/mission: STS-6
Crew: Paul Weitz, Karol Bobko, Donald Peterson, Story Musgrave
Spacewalkers: Donald Peterson, Story Musgrave
Purpose: Test STS EMU and EVA equipment

The first Shuttle EVA (and first U.S. EVA since February 1974) was planned to occur during STS-5, but was scrubbed because of Shuttle EMU malfunctions. Story Musgrave checked out the three suits - two primary and one backup - carried in Challenger's airlock early in the flight to provide additional time for troubleshooting should one of the 125 kg (275 lb) suits prove faulty. The Shuttle EMU is first operational U.S. space suit built specifically for EVA. Nominal suit operating pressure is 29.7 kilopascal (4.3 psi). The EMU consists of the Primary Life Support System (PLSS) and the Space Suit Assembly (SSA). The SSA is built around the fiberglass Hard Upper Torso (HUT). Water, oxygen, electricity, and data pass between the PLSS and the HUT through an interface pad behind the astronaut's left shoulder. Most SSA components can fit men and women from the 5th to 95th percentiles of body size. There are four HUT sizes, six waist bearing sizes, and two boot sizes (the latter with six sizes of sizing insert "slippers"). There are also nine standard glove sizes, but generally astronauts opt for customized gloves when possible. This is the only customizable part of the Shuttle EMU, pointing up the importance placed on adequate gloves in EVA work. The Display and Control Module (DCM) is mounted on the front of the HUT. A microprocessor in the DCM monitors suit condition - the EMU is the first computerized space suit. The astronauts reported after the flight that, despite attempts by engineers to make the Shuttle EMU self-donning, an IV crewman was required to close the EMU waist ring. The exercise treadmill and extra EMU interfered with suit donning. Musgrave and Peterson "stood" face to face in the airlock, which lead to thrashing and noise. Despite this, they managed to doze during the 3-hr in-airlock prebreathe. The astronauts left the airlock over Borneo and assessed EMU mobility by translating aft along the handholds inside the payload bay door hinge. STS-6 CDR and Skylab 2 EVA veteran Paul Weitz and Pilot Karol Bobko observed from the aft flight

STS-6, 1983 - Donald Peterson uses handrails and a tether slidewire to translate down Challenger's payload bay door sill toward Story Musgrave during the first Space Shuttle EVA. (S06-44-0549)

deck. Musgrave climbed to the top of the payload bay aft bulkhead and looked back over Challenger's engine bells. He referred to his Weightless Environmental Training Facility (WETF) neutral buoyancy EVA training when he quipped that "this is a little deeper pool than I'm used to working in." Shuttle Program Manager Glynn Lunney said that "EVA on Shuttle is wide-ranging even when you stay in the cargo bay." Musgrave and Peterson demonstrated contingency payload bay door closure without actually closing Challenger's 18.3-m-long (60-ft-long) doors. They had difficulties rewinding the payload bay door EVA winch and considered cutting the winch cable, but Mission Control vetoed this and the cable came free. They also went through the motions of lowering a jammed satellite tilt table. Musgrave reported that his fingers were cold. According to NASA Associate Administrator for Space Flight James Abrahamson, the STS-6 mission "was flown exactly as planned, including a marvelous EVA."

Memorandum CF-4, STS-6 Crew Debriefing, Appendix VII, April 25, 1983; "Mission 6 EVA Clears Way for Untethered Operations," *Aviation Week & Space Technology*, April 11, 1983, p. 25; "F-Troop Adds New Spacecraft to the Fleet," *Space News Roundup*, April 13, 1983, p. 1; *Astronautics and Aeronautics, 1979-1984*, NASA SP 4024, 1990, pp. 407-408; *Systems Division EVA Prep/Post Training Workbook*, JSC-23901, Rev. A, November 1989; *Extravehicular Mobility Unit Systems Training Workbook*, JSC 19450, Mission Operations Directorate, September 1989; *EVA Contingency Operations Training Workbook*, 88-054, Rev A, Mission Operations Directorate, March 1995.

April 9	**STS-6/Challenger landing**
April 20-22	**Salyut 7/Soyuz-T 8**
June 18-24	**STS-7/Challenger**
June 27	**Salyut 7/Soyuz-T 9 PE-2 launch**
August 30-September 5	**STS-8/Challenger**
September 26	**Salyut 7/Soyuz-T 10a**

November 1
1983 EVA 2
World EVA 46
Russian EVA 7
Space Station EVA 15
Duration: 2:50
Spacecraft/mission: Salyut 7 PE-2
Crew: Vladimir Lyakhov, Alexandr Alexandrov
Spacewalkers: Vladimir Lyakhov, Alexandr Alexandrov
Purpose: Augment Salyut 7 solar array

The solar arrays on the Salyut 6 space station underwent rapid degradation in their ability to produce electrical power due to UV and atomic oxygen exposure, so Salyut 7 was designed to have its arrays augmented over the course of its occupancy to restore lost capacity. Solar array extensions for the central (top) array were delivered by the automated Cosmos 1443 vehicle (March 1983). The EVA to augment the power-starved station's solar arrays was originally to have been performed by cosmonauts Vladimir Titov and Gennadi Strekalov, but they were unable to dock their Soyuz T-8 spacecraft with Salyut 7. Strekalov and Titov's next attempt to pay a service call to Salyut 7 (Soyuz-T 10a) was stymied when their launch vehicle exploded seconds before scheduled liftoff. The Soyuz-T escape system functioned as designed, so neither Strekalov nor Titov was injured. However, the Salyut 7 power shortage grew acute, so flight controllers tapped Lyakhov and Alexandrov to perform the EVA. Some sources state that the cosmonauts practiced the procedure a dozen times in the Hydrolaboratory at Star City. During preparations, Alexandrov discovered a tear in his suit's primary pressure bladder. The cosmonauts mended this and the suit performed nominally during the EVA. For this EVA the cosmonauts used the same Orlan-D space suits that Lebedev and Berezevoi used. Lyakhov became the first Russian to perform a second EVA. Alexandrov set up a TV camera on a movable arm so flight controllers in the TsUP could monitor the EVA, then the cosmonauts took up position in foot restraints and removed the add-on array from its container. Salyut 6 EVA cosmonaut Valeri Ryumin was shift chief in the TsUP for the EVA. Forty min into the EVA they passed out of range of Soviet ground stations and tracking ships for 50 min. Much of the time out of range the cosmonauts spent in darkness. They awaited orbital sunrise so they could resume work. They then used a special "compact and convenient" winch to unfurl the add-on array along one side of the existing array. A total of 48 operations were needed to deploy the array nominally, with up to 189 operations required in contingency situations. The 5-m-long (16.4-m-long), 1.5-m-wide (5-ft-wide) add-on array increased available power by 25 percent. Lyakhov received a reprimand from the TsUP for releasing bits of junk to watch them float away - the glittering objects could interfere with Salyut 7's star sensors.

"Above the Planet: Salyut EVA Operations," Neville Kidger, *Spaceflight*, March 1989, pp. 105; "Salyut 7 - Our Commentary - Space Installers," V. Vladimirov, *Pravda*, November 4, 1983, p. 3 (translated in *USSR Report: Space*, JPRS-USP-84-002, March 16, 1984, pp. 16-18); "The Experience in Operation and Improving the Orlan-type Space Suits," I. P. Abramov, *Acta Astronautica*, Vol. 36, No. 1, July 1995, pp. 1-12.

November 3
1983 EVA 3
World EVA 47
Russian EVA 8
Space Station EVA 16
Duration: 2:55
Spacecraft/mission: Salyut 7 PE-2
Crew: Vladimir Lyakhov, Alexandr Alexandrov
Spacewalkers: Vladimir Lyakhov, Alexandr Alexandrov
Purpose: Augment Salyut 7 solar array

During this EVA, Lyakhov and Alexandrov followed the same procedure they used to install the first add-on solar array two days earlier. This marked the first time the Soviets performed two EVAs in one mission. The EVA was planned so that the principal activities occurred when the station was in daylight and in radio contact with the TsUP. The two add-on arrays installed by Lyakhov and Alexandrov increased Salyut 7's electrical capacity by 800W. According to Viktor Blagov, Deputy Flight Director at the TsUP, the EVAs were

> . . .important for two reasons. . . solar battery elements gradually lose their productivity when they are operated in space for a very long time. . . [while at the same time] instruments which require more and more energy are being sent into orbit. In the future we will. . . attach special scientific modules which also require a great deal of energy. And today's installation operations are the very first steps in solving the orbital energy problems facing us. . . in the second place, although all kinds of maintenance work has been carried out in open space before, up to now such major installation operations have not been conducted. After all, the time is not far off when brigades of installation workers will fly into orbit and build large orbital complexes in space. [Lyakhov and Alexandrov] have already proved that all this is completely feasible.

Leonid Kizim and Vladimir Solovyov simultaneously simulated the work in the Hydrolaboratory neutral buoyancy facility in Star City. In a 1984 article, Alexandrov and Lyakhov described them as "the test pilots who trained us" and said that "in the event of difficulties the duplicating team on the ground could give us assistance with their recommendations."

"Above the Planet: Salyut EVA Operations," Neville Kidger, *Spaceflight*, March 1989, pp. 105; "Above the Abyss Once Again - Report from the Flight Control Center," A. Ivanokhov, *Izvestia*, November 4, 1983, p. 2 (translated in *USSR Report: Space*, JPRS-USP-84-002, March 16, 1984, pp. 13-15); *Zemlya i Vselennaya*, Vladimir Lyakhov and Alexandr Alexandrov, May-June 1984, pp. 5-11 (translated in *USSR Report: Space*, JPRS-USP-85-001, February 4, 1985, pp. 16-17).

November 23   **Salyut 7/Soyuz-T 9 PE-2 landing**

November 29-December 8   **STS-9/Columbia**

1984

February 3   **STS-41B/Challenger launch**

February 7
1984 EVA 1
World EVA 48
U.S. EVA 40
Shuttle EVA 2
MMU EVA 1
Duration: 5:55
Spacecraft/mission: STS 41-B
Crew: Vance Brand, Robert Gibson, Bruce McCandless, Robert Stewart, Ronald McNair
Spacewalkers: Bruce McCandless, Robert Stewart
Purpose: MMU tests; SMM repair rehearsal

"It may have been one small step for Neil but it's a heck of a big leap for me," quipped astronaut Bruce McCandless, MMU co-designer - with Charles Whitsett of NASA's Johnson Space Center (JSC) - as he took the Martin-Marietta Manned Maneuvering Unit (MMU) on its first test flight. The twin flight MMUs were delivered to JSC in Houston for acceptance testing in September 1983. The first MMU EVA began on mission day 5 of STS 41-B. For the EVA, Challenger's CDR Vance Brand provided intravehicular support during suit-up, monitored MMU speed and distance using Challenger's radar, and flew the orbiter, while Ronald McNair operated the Remote Manipulator System (RMS) mechanical arm. Jerry Ross was EVA CapCom in Houston for this and all subsequent MMU EVAs. In addition to testing the MMU in flight, McCandless and fellow MMU flyer Robert Stewart performed a dress rehearsal for the Solar Maximum Mission (Solar Max) satellite retrieval, scheduled for the next flight (STS 41-C). Both flight MMUs were carried in the forward part of Challenger's payload bay. To help relieve the thrashing and banging which occurred during airlock depressurization on STS-6, McCandless pointed his head toward the airlock floor while Stewart pointed toward the ceiling. McCandless spent 90 min checking and donning the port MMU, then tested it in the payload bay by maneuvering precisely around equipment. He found that the backpack shuddered and shook when forward movement was initiated in attitude hold. McCandless then moved 45 m (150 ft) out from Challenger, returned to the payload bay, flew out to 96 m (315 ft) and returned, then moved out again to about 99 m (325 ft). MMU nitrogen propellant use was higher than in simulations. Brand noted that the MMU's tracking lights were inadequate for finding the astronaut if he strayed away during orbital night, so ordered McCandless to hurry back to the payload bay before Challenger passed into darkness. Meanwhile, Stewart installed a Manipulator Foot Restraint (MFR) on the RMS, but had to postpone a test ride because the EVA was behind schedule. When McCandless returned to the payload bay, Stewart attached between the MMU arms the Trunnion Pin Attachment Device

(TPAD) to be used to snare Solar Max. McCandless practiced docking with a trunnion pin mounted next to a mockup of the Solar Max main electronics box in the payload bay. McCandless reported later that he was chilled when out away from the payload bay. Stewart then flew the MMU 93 m (306 ft) from Challenger. Brand noted that Stewart was traveling at 0.6 mps (2 fps) about 90 m (300 ft) from Challenger, so warned him to slow down. Stewart tested the MMU for 65 min. Meanwhile, McCandless became the first astronaut to ride the MFR at end of RMS. The arm proved more stable for EVA work than expected.

STS 41-B Pilot's Report, June 1984; "Astronauts Evaluate Maneuvering Backpacks," Craig Covault, *Aviation Week & Space Technology*, February 13, 1984, pp. 16-19; "Steppin' Out with Flash and Buck," *Space News Roundup*, NASA JSC, February 24, 1984, pp. 1-2.

STS 41-B, 1984 - Bruce McCandless (pictured here) and Robert Stewart tested the MMU outside Challenger. (S84-27562)

February 8   Salyut 7/Soyuz-T 10b PE-3 launch

February 9
1984 EVA 2
World EVA 49
U.S. EVA 41
Shuttle EVA 3
MMU EVA 2
Duration: 6:17
Spacecraft/mission: STS 41-B
Crew: Vance Brand, Robert Gibson, Bruce McCandless, Robert Stewart, Ronald McNair
Spacewalkers: Bruce McCandless, Robert Stewart
Purpose: Test MMU; rehearse Solar Max repair

On mission day 7 Stewart and McCandless ventured outside again, this time with Pilot Robert Gibson operating the RMS. The manipulator arm suffered wrist joint and elbow TV camera malfunctions prior to this EVA. The camera problem meant that engineering film coverage of the EVA was not as comprehensive as planned. The joint failure was much more serious - it meant that the RMS could not release a rotating Shuttle Pallet Satellite above Challenger's payload bay, depriving the MMU astronauts of a spinning target for practice docking using the TPAD. The ability to dock with a rotating target was considered crucial to the Solar Max repair scheduled for STS 41-C. The astronauts practiced docking with fixed targets instead. Stewart performed a hydrazine transfer experiment to help validate the orbiter's proposed role as a satellite tanker. Freon dyed red for visibility filled in for poisonous hydrazine. A foot restraint worked itself loose; Brand maneuvered Challenger and McCandless moved down the starboard sill to retrieve the errant hardware. NASA called this an unplanned test of the Shuttle's ability to rescue an astronaut stranded by MMU failure.

STS 41-B Pilot's Report, June 1984; "Astronauts Evaluate Maneuvering Backpacks," Craig Covault, *Aviation Week & Space Technology*, February 13, 1984, pp. 16-19; "Steppin' Out with Flash and Buck," *Space News Roundup*, NASA JSC, February 24, 1984, pp. 1-2.

February 11   STS-41B/Challenger landing

April 3   Salyut 7/Soyuz-T 11 VE-3 launch

April 6   STS-41C/Challenger launch

April 8
1984 EVA 3
World EVA 50
U.S. EVA 42
Shuttle EVA 4
MMU EVA 3
Duration: 2:38
Spacecraft/mission: STS 41-C
Crew: Robert Crippen, Francis Scobee, Terry Hart, James van Hoften, George Nelson
Spacewalkers: George Nelson, James van Hoften
Purpose: Retrieve Solar Max satellite

Challenger launched within a tight window to rendezvous with Solar Max. NASA considered the mission a critical demonstration of the Shuttle's ability to service satellites. As the orbiter approached the science satellite, the crew reduced cabin pressure to 70.3 kpascal (10.2 psi) to minimize EVA prebreathe time. On this date Pilot Dick Scobee helped Nelson and Van Hoften don their EMUs. After entering Challenger's payload bay, the spacewalkers discovered that the payload bay door slidewires were looser than expected. Nelson donned the MMU, then attempted to dock with Solar Max using the TPAD mounted between the hand controller arms. He bounced off Solar Max after the TPAD jaws failed to close on one of the satellite's berthing docking pins. Solar Max began to spin. Twice more he attempted to latch onto the satellite with the TPAD, each time adding to the slow spin. His difficulties were later traced to an obstructing grommet on Solar Max which did not appear in its blueprints. Nelson then tried to stabilize the satellite by gripping one of its two solar arrays and activating the MMU's automatic attitude hold feature, but this reversed the spin and started an unpredictable tumble about two axes. Solar Max lost its lock on the Sun and began draining its batteries. Nelson was forced to return to Challenger when his MMU nitrogen propellant supply ran low. MMU co-designer Charles Whitsett said later that the MMU's operating temperature was low, reducing nitrogen pressure, so there was probably more

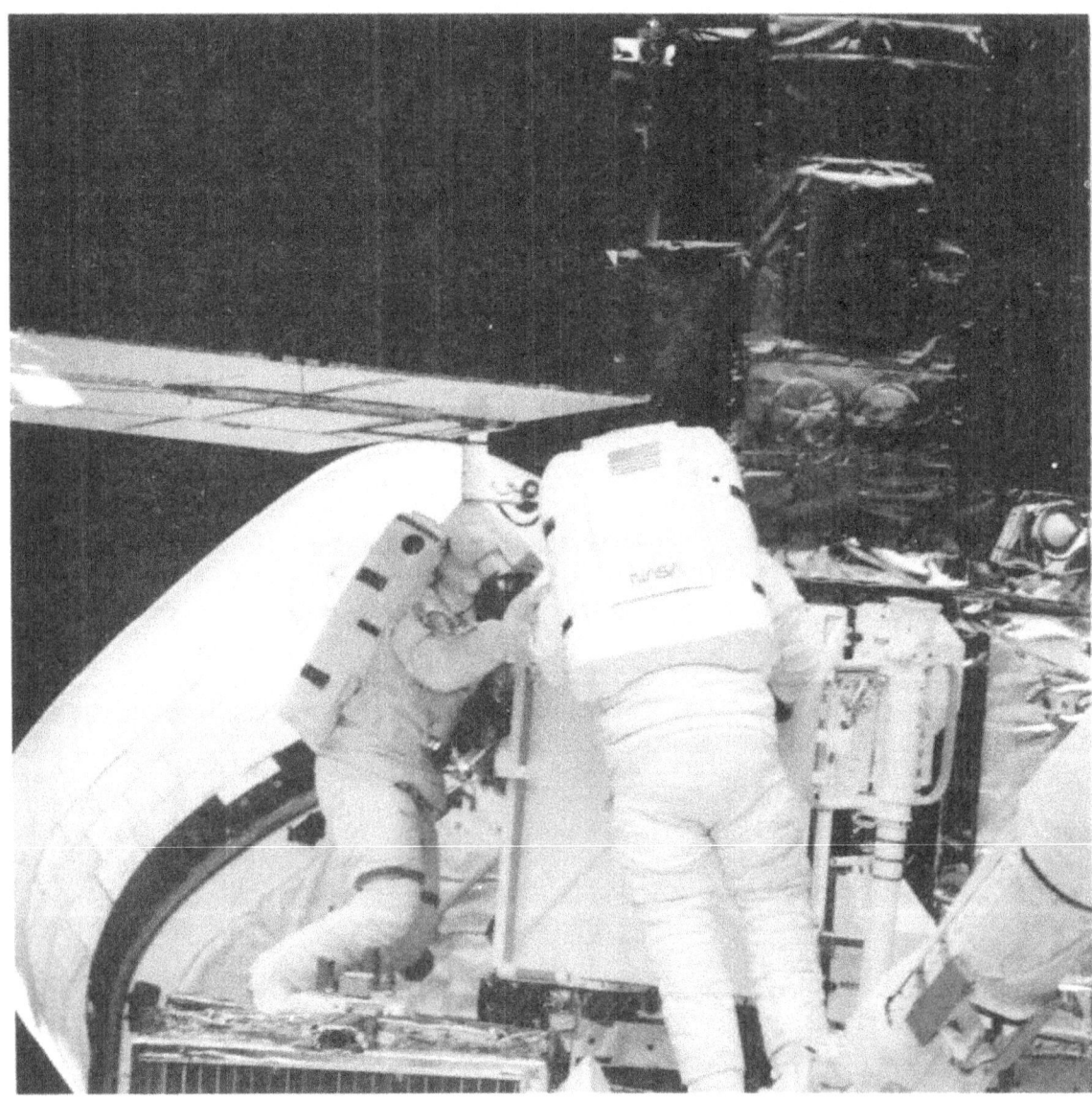

STS 41-C, 1984 - George Nelson (left) and James van Hoften service the Solar Maximum Mission satellite (top right) in Challenger's payload bay. (41C-38-1853)

nitrogen available than Nelson thought - perhaps enough for another stabilization attempt. The astronauts and Mission Control considered changing MMUs and TPADs and trying again, but Challenger's rendezvous fuel was running low, threatening the orbiter's ability to recover a stranded MMU astronaut. A subsequent attempt by Terry Hart to capture the satellite using the RMS failed because of the tumble. The astronauts returned to the airlock. According to an *Aviation Week* editorial, the evening of April 8 "seemed like ebb tide for shuttle credibility. It had failed to live up to one of its development justifications: payload retrieval and refurbishment. A prospective Palapa communications satellite rescue attempt was beginning to seem like a grim joke. A Landsat 4 rescue in 1986, which needs money and work to start now, looked no better."

STS 41-C Flight Crew Report, (no date); "Black Sunday. . . Fat Tuesday," William Gregory, *Aviation Week & Space Technology*, April 16, 1984, p. 13; "Orbiter Crew Restores Solar Max," Craig Covault, *Aviation Week & Space Technology*, April 16, 1984, pp. 18-20; "NASA Believes EVAs Valid Despite Recovery Problem," *Aviation Week & Space Technology*, April 1984, pp. 21-24; "Repair Mission Became High Drama," *Space News Roundup*, NASA JSC, April 27, 1984, pp. 1-2.

April 11
1984 EVA 4
World EVA 51
U.S. EVA 43
Shuttle EVA 5
MMU EVA 4
Duration: 6:44
Spacecraft/mission: STS 41-C
Crew: Robert Crippen, Francis Scobee, Terry Hart, James van Hoften, George Nelson
Spacewalkers: George Nelson, James van Hoften
Purpose: Repair Solar Max satellite

NASA Goddard Space Flight Center (GSFC), which operated Solar Max, stopped its tumble on April 9 before its batteries became depleted. On April 10, Terry Hart grappled the satellite with the RMS on the first try while Crippen piloted Challenger. Hart placed the satellite in its servicing cradle in the payload bay. Van Hoften and Nelson entered the bay on this date for an EVA scheduled to last 6 hr. They completed replacement of the satellite's 227-kg (500-lb) attitude control and main electronics box 1 hr ahead of schedule. Solar Max was a NASA GSFC-developed Multimission Modular Spacecraft (MMS) designed for routine servicing. Van Hoften reported that the EVA "tools are working great - haven't had one glitch yet." They praised the Module Servicing Tool, which was developed specifically for MMS servicing. The failed attitude control module and main electronics box of the satellite were stowed for return to Earth, where they would be analyzed to determine the cost-effectiveness of satellite refurbishment in orbit. If only certain satellite components wore out, then it would be cost-effective to make occasional repairs and change out instruments; if the entire satellite degraded at the same rate, it would be less costly to launch a new satellite. They also stowed thermal blankets and various aluminum parts for analysis by orbital debris researchers on Earth. The salvaged components acted as impromptu debris catchers during their 4 yr in space. Nelson experienced a minor urine containment problem; the liquid cooling garment absorbed most of the leakage, so the largest impact was on post-EVA cleanup. He noted some helmet fogging, though postflight inspection showed that no urine migrated to his helmet through the air circulation system. The fogging occurred when Nelson turned down the flow of cooling water to his Liquid Cooling Ventilation Garment (LCVG) after he became too cold. This reduced ventilation and allowed condensation to build up on the inside of his faceplate. Before returning to the airlock, Nelson stepped into an MFR and rode the RMS above Solar Max to examine and photograph the satellite from all angles. Van

Hoften then took the MMU on a short test flight in the payload bay. Frank Cepollina, Solar Max Recovery Mission Project Manager at NASA GSFC, told the astronauts and other NASA JSC employees and contractors involved in the 3-yr effort to rescue the satellite that, "The engineers and managers at Goddard have asked me to express to all of you their respect and admiration for your efforts, and appreciation for a job well done."

STS 41-C Flight Crew Report, (no date); "USAF, NASA Discuss Shuttle Use for Satellite Maintenance," Craig Covault, *Aviation Week & Space Technology*, December 17, 1984, pp. 14-17; *Astronautics and Aeronautics. 1979-1985*, NASA, 1990, p. 474; "Black Sunday. . . Fat Tuesday," William Gregory, *Aviation Week & Space Technology*, April 16, 1984, p. 13; "Orbiter Crew Restores Solar Max," Craig Covault, *Aviation Week & Space Technology*, April 16, 1984, pp. 18-20; "NASA Believes EVAs Valid Despite Recovery Problem," *Aviation Week & Space Technology*, April 1984, pp. 21-24; "Repair Mission Became High Drama," *Space News Roundup*, NASA JSC, April 27, 1984, pp. 1-2.

April 11   **Salyut 7/Soyuz-T 10b VE-3 landing**

April 13   **STS-41C/Challenger landing**

April 23
1984 EVA 5
World EVA 52
Russian EVA 9
Space Station EVA 17
Duration: 4:20
Spacecraft/mission: Salyut 7 PE-3
Crew: Leonid Kizim, Vladimir Solovyov, Oleg Atkov
Spacewalkers: Leonid Kizim, Vladimir Solovyov
Purpose: Prepare worksite for ODU repair

On April 17 Progress 20 docked at Salyut 7's aft port carrying 25 tools and other equipment for repairing the station's main propellant system, the ODU, which suffered an oxidizer system rupture on September 9, 1983. The ODU was located in Salyut 7's unpressurized aft equipment compartment. There were no handholds near the ODU, so engineers attached a special work platform with foot restraints for holding the cosmonauts at the worksite to Progress 20's forward dry cargo module. The TsUP extended the platform by remote control prior to the EVA. Atkov provided support from inside Salyut 7. Total distance between the Salyut 7 airlock hatch and the worksite was 15 m (49 ft). The cosmonauts' progress over the station's hull was impeded by the 40 kg (88 lb) of equipment they carried; this included a tool caddy, cutting tools, wrenches, bypass pipes, a waste container, and, most encumbering of all, a ladder for reaching the worksite. Kizim and Solovyov drove anchor pins into the equipment compartment's plastic skin to attach the ladder and tool containers, then unfolded the ladder to its full 5-m (16-ft) length before closing out this first of two EVAs scheduled for April.

Pravda, April 24, 1984, p. 1 (translated in *USSR Report: Space*, JPRS-USP-84-006-L, July 20, 1984, p. 31); "Above the Planet: Salyut EVA Operations" (Part Two), Neville Kidger, *Spaceflight*, March 1989, pp. 105; *The Soviet Year in Space 1984*, Nicholas Johnson, 1985, pp. 40-42; *Gudok*, April 27, 1984, p. 3 (translated in *USSR Report: Space*, JPRS-USP-84-006-L, July 20, 1984, p. 33); *Kosomolskaya Pravda*, April 27, 1984, p. 4 (translated in *USSR Report: Space*, JPRS-USP-84-006-L, July 20, 1984, p. 34).

April 26
1984 EVA 6
World EVA 53

Russian EVA 10
Space Station EVA 18
Duration: 4:56
Spacecraft/mission: Salyut 7 PE-3
Crew: Leonid Kizim, Vladimir Solovyov, Oleg Atkov
Spacewalkers: Leonid Kizim, Vladimir Solovyov
Purpose: Repair Salyut 7 ODU

Salyut 6 EVA veteran Valeri Ryumin monitored Kizim and Solovyov from the TsUP as they performed their second ODU repair EVA, which was scheduled to last 4 hrs, 5 min. The EVA occurs in the early morning hours, Moscow time. Salyut 7's orbital geometry meant that radio relay ships in the Atlantic and Pacific permitted 20-50 min of communication between station and TsUP during each 90-min orbit. Kizim and Solovyov needed about 20 min to move from the Salyut 7 airlock to the worksite, where they set up a TV camera so the TsUP could watch their work. Kizim took up position on the ladder installed on the previous EVA, while Solovyov placed his boots in the foot restraints on the Progress 20 extension. They pulled aside thermal blankets and cut through the station's plastic skin to reach the oxidizer plumbing. The cosmonauts located and replaced a valve on a "shut-off part of the reserve line," but only after a nut locked by epoxy resin thwarted their efforts for 2 hr. The oxidizer system was then pressurized with nitrogen to check their work, revealing that the ODU still leaked. The cosmonauts asked for and received an extension to complete work on the reserve line; when the extension lapsed Ryumin had to order the cosmonauts back inside Salyut 7.

"Above the Planet: Salyut EVA Operations (Part Two)," Neville Kidger, *Spaceflight*, March 1989, pp. 140; *Sovetskaya Latviya*, April 28, 1984, p. 1 (translated in *USSR Report: Space*, JPRS-USP-84-006-L, July 20, 1984, p. 32); *Gudok*, April 27, 1984, p. 3 (translated in *USSR Report: Space*, JPRS-USP-84-006-L, July 20, 1984, p. 33); *Kosomolskaya Pravda*, April 27, 1984, p. 4 (translated in *USSR Report: Space*, JPRS-USP-84-006-L, July 20, 1984, p. 34).

April 29
1984 EVA 7
World EVA 54
Russian EVA 11
Space Station EVA 19
Duration: 2:45
Spacecraft/mission: Salyut 7 PE-3
Crew: Leonid Kizim, Vladimir Solovyov, Oleg Atkov
Spacewalkers: Leonid Kizim, Vladimir Solovyov
Purpose: Repair Salyut 7 ODU

Kizim and Solovyov lobbied for a third April EVA and, after some debate on the ground, were granted permission to attempt to complete the ODU repair. With this EVA Kizim and Solovyov became the first Soviet spacewalkers to conduct three EVAs in one flight, an honor first enjoyed in 1966 by U.S. astronaut Edwin Aldrin on Gemini 12. Atkov monitored crew status from inside Salyut 7. The cosmonauts finished work on the line they repaired during their second EVA, then installed a bypass line between two fill tubes, creating a new conduit to the main oxidizer supply. After they completed their work, nitrogen was again pumped through the system to check its integrity. To the dismay of all, the ODU plumbing still leaked. Kizim and Solovyov replaced the thermal blankets and returned inside while troubleshooters on the ground resumed efforts to localize the leak.

"Above the Planet: Salyut EVA Operations (Part Two)," Neville Kidger, *Spaceflight*, March 1989, pp. 140;

Moskovskaya Pravda, April 30, 1984, p. 3 (translated in *USSR Report: Space*, JPRS-USP-84-006-L, July 20, 1984, p. 35).

May 4
1984 EVA 8
World EVA 55
Russian EVA 12
Space Station EVA 20
Duration: 2:45
Spacecraft/mission: Salyut 7 PE-3
Crew: Leonid Kizim, Vladimir Solovyov, Oleg Atkov
Spacewalkers: Leonid Kizim, Vladimir Solovyov
Purpose: Repair Salyut 7 ODU

The fourth Salyut 7 PE-3 EVA occurred on the cosmonauts' 85th day in space. According to Solovyov, by this time they were adept at moving over Salyut 7's hull. The cosmonauts removed the thermal blankets again and installed a second conduit in the Salyut 7 oxidizer system. Atkov and controllers in the TsUP were then able at last to pin down the precise location of the ruptured pipe. Kizim and Solovyov were dismayed to learn that they lacked tools adequate to complete the repair. They replaced the thermal blankets and rejoined Atkov inside Salyut 7, their efforts again thwarted. Progress 20 undocked on May 6, taking the special extension and foot restraints with it. The cargo ship reentered Earth's atmosphere and burned up on May 7.

"Above the Planet: Salyut EVA Operations (Part Two)," Neville Kidger, *Spaceflight*, March 1989, pp. 140; *Izvestiya*, May 5, 1984, p. 2 (translated in *USSR Report: Space*, JPRS-USP-84-006-L, July 20, 1984, p. 37); "Salyut 7: Third Expedition to the Station," S.A. Bovin, *Zemlya i Vselennaya*, March-April 1985, pp. 9-15 (translated in *USSR Report*: Space, JPRS-USP-86-001, January 13, 1986, pp. 57-58).

May 19
1984 EVA 9
World EVA 56
Russian EVA 13
Space Station EVA 21
Duration: 3:05
Spacecraft/mission: Salyut 7 PE-3
Crew: Leonid Kizim, Vladimir Solovyov, Oleg Atkov
Spacewalkers: Leonid Kizim, Vladimir Solovyov
Purpose: Augment Salyut 7 solar array

Salyut 7 had three solar arrays at launch, all of which were scheduled to be augmented over the period of the station's occupancy. Augmenting the center (top) array required two EVAs in November 1983. The Progress 21 automated freighter delivered extensions for the port array on May 10. During this period Progress flights were frequent to make up for air spilled into space from the transfer compartment during EVAs, and because of the added logistics requirements of having three crew members on board Salyut 7. Adding the port array extensions required only one EVA, demonstrating the benefits of EVA experience (this was Kizim and Solovyov's fifth EVA) and of applying lessons learned from the November EVAs, during which Kizim and Solovyov performed neutral buoyancy simulations of the array installation in the Star City Hydrolaboratory. The new panels contained cells made of gallium arsenide that were more efficient at producing electricity than the silicon cells launched with Salyut 7. Kizim and Solovyov left the airlock toting tools and the two solar array extensions, each in a separate container. They discarded the containers after removing the panels, taking "care. . . to cast them into a different orbit, to prevent the station from encountering them in the future." From foot restraints Solovyov and Kizim assembled each 4.56-sq-m (49-sq-ft) extension, then attached the first and winched it into position. Atkov used controls inside Salyut 7 to turn the port array 180 deg so it presented its other side to the cosmonauts, who then attached and winched into place the second add-on panel. Solovyov struggled to tie two knots in wire bundles linking the arrays to the station's main external power panel, a task he later compared to "trying to thread a needle in boxing gloves." Their work added 1.2kW to Salyut 7's power supply. With this EVA, their fifth together, Kizim and Solovyov tied David Scott's record for total career EVAs.

Pravda, May 20, 1984, p. 1 (translated in *USSR Report: Space*, JPRS-USP-84-006-L, July 20, 1984, p. 44); "Above the Planet: Salyut EVA Operations (Part Two)," Neville Kidger, *Spaceflight*, March 1989, pp. 140; *Trud*, May 20, 1984, p. 3 (translated in *USSR Report: Space*, JPRS-USP-84-005, October 26, 1984, p. 3); *Pravda*, June 3, 1984, p. 3 (abstracted in *USSR Report: Space*, JPRS-USP-84-005, October 26, 1984, p. 9).

July 17   **Salyut 7/Soyuz-T 12 VE-4 launch**

July 25
1984 EVA 10
World EVA 57
Russian EVA 14
Space Station EVA 22
Duration: 3:35
Spacecraft/mission: Salyut 7 VE-4
Crew: Vladimir Dzhanibekov, Svetlana Savitskaya, Igor Volk (VE-4); Leonid Kizim, Vladimir Solovyov, Oleg Atkov (PE-3)
Spacewalkers: Svetlana Savitskaya, Vladimir Dzhanibekov
Purpose: Perform first EVA by a woman; test the URI electron beam tool

For this first EVA by a woman, Savitskaya donned an Orlan-D suit already worn eight times by cosmonauts on Salyut 7. With Dzhanibekov, she was tasked with testing the *Universalny Rabochy* (or *Ruchnoj) Instrument* ("Universal Hand Tool") (URI) multipurpose electron beam cutting, welding, soldering, and brazing tool. Savitskaya played a central role in developing the handle and other cosmonaut interfaces of the tool. She trained with URI three times in a vacuum chamber and in a plane flying parabolas. Some engineers voiced reservations about flying URI - it generated a great deal of heat which might damage the cosmonauts' space suits. The experience of the Vulkan automated welding system 15 yr before loomed large in engineers' minds (the device ran amok aboard Soyuz 6 and nearly cut the table holding welding samples in half). On day 7 of VE-4, with Igor Volk inside Salyut 7 monitoring the EVA timeline, Dzhanibekov opened the Salyut 7 airlock. He unfolded and stood in a Yakor foot restraint, then set up a worksite lamp. Savitskaya handed out URI, which Dzhanibekov set up and attached to an external power outlet. He then traded places with Savitskaya, who set up a TV camera. Salyut 7 passed out of communications range with the TsUP; when contact was restored, Savitskaya began work with URI, first cutting a 0.5-mm- (0.02-in-) thick titanium sample. In all she performed six cutting, two silver spray coating, and six soldering experiments, taking care always not to point URI at Salyut 7 lest the tool run amok. Her heart rate peaked during the EVA at 140 beats/min. While soldering the Sun glared in her face, making it difficult for her to see her work; nevertheless, her results were later judged satisfactory. Savitskaya and Dzhanibekov then traded places again so he could test URI. Dzhanibekov said later that "the tool is very handy and I'm sure we'll be using it a lot." After finishing, he took down URI and handed the device and experiment samples to Savitskaya. Dzhanibekov then removed Ekpozitsiya cassettes from the station's exterior and handed them to Savitskaya, who handed back a Meduza bio-polymer cassette for installation. Products of the welding experiment returned to Earth in Soyuz T-12. *Kosomolskaya Pravda* reported on the EVA, saying that: "The world's first egress into open space by a woman cosmonaut has been made by Svetlana Savitskaya. Her successful performance of unique experiments in conditions of outer space demonstrated that it is possible for a woman to function effectively while performing complex work not only on board a manned orbiting complex but also in open space." In January 1985, Boris Paton referred to Savitskaya's experiments, saying that "the time when robots will be serving as welders in outer space is not too distant."

"Above the Planet: Salyut EVA Operations (Part Two)," Neville Kidger, *Spaceflight*, March 1989, pp. 140-141; *Kosomolskaya Pravda*, July 27, 1984, p. 1 (translated in *USSR Report: Space*, JPRS-USP-84-005, October 26, 1984, p. 39); *Sotsialisticheskaya Industriya*, July 27, 1984, p. 1 (abstracted in *USSR Report: Space*, JPRS-USP-84-005, October 26, 1984, p. 40); *Izvestia*, July 27, 1984, p. 3 (abstracted in *USSR Report: Space*, JPRS-USP-84-005, October 26, 1984, p. 41); *Avtomaticheskaya Svarka*, June 1986, pp. 1-4 (translated in *USSR Report: Space*, JPRS-USP-86-008-L, December 16, 1986, pp. 43-51); "Salyut 7: Third Expedition to the Station," S.A. Bovin, *Zemlya i Vselennaya*, March-April 1985, pp. 9-15 (translated in *USSR Report*: Space, JPRS-USP-86-001, January 13, 1986, pp. 58-59); "The Mir Complex, Our Commentary: A Bridge to the Stars Themselves," A. Tarasov, *Pravda*, March 29, 1988, p. 2 (translated in *JPRS Report, Science & Technology, USSR: Space*, August 17, 1988, pp. 46-49); "Soviets Resupply Salyut, Provide EVA Details," *Aviation Week & Space Technology*, August 20, 1984, p. 25.

August 8
1984 EVA 11
World EVA 58
Russian EVA 15
Space Station EVA 23
Duration: 5:00
Spacecraft/mission: Salyut 7 PE-3
Crew: Leonid Kizim, Vladimir Solovyov, Oleg Atkov
Spacewalkers: Leonid Kizim, Vladimir Solovyov
Purpose: Complete Salyut 7 propulsion system repair

The VE-4 crew delivered a Portable Pneumo Press to the PE-3 crew of Kizim, Solovyov, and Atkov. The tool was developed specifically to allow completion of the Salyut 7 ODU oxidizer system repair. Dzhanibekov received training in its use on Earth, and in turn trained Kizim and Solovyov aboard Salyut 7 during his visit. VE-4 also delivered an instructional videotape, manuals, and photos of the device in operation. Kizim and Solovyov moved to the worksite, pulled back the thermal blankets, and used the press to squeeze a stainless steel pipe. Checks showed that the ODU oxidizer system was at last sealed. With this EVA, their sixth together, Kizim and Solovyov broke David Scott's 1971 record for total career EVAs and completed a record 22 hr. 50 min of EVA in a single mission. The EVA also marked the tenth EVA for the Orlan-D suits they wore and the last use of the Orlan-D suit. Perhaps because it was old, Solovyov's suit suffered cooling water pump failure during the EVA. He compensated by operating primary and backup circulating fans simultaneously and resting periodically to cool off. Before returning inside, Kizim and Solovyov removed a sample of silicon solar cell material so engineers could study its degradation. They used a special holding tool to avoid contaminating it with their suit gloves. Physician Atkov later reported that the men's hands were in bad shape after the EVA, "as if they had been in a fist fight," though their general health remained good. In an interview after the flight, the cosmonauts said that their work was "a rather good rehearsal for future major installations," and that "in future such work will be indispensable for the servicing of. . . satellite systems." "The experience," they added, "is also valuable in that it forces the crew to learn 'on the move,' when already aboard the space station."

"Above the Planet: Salyut EVA Operations," Neville Kidger, *Spaceflight*, May 1989, p. 154; *Krasnaya Zvezda*, August 10, 1984, p. 1 (translated in *USSR Report: Space*, JPRS-USP-84-006, November 14, 1984,

p. 5); *Pravda*, August 9, 1984, p. 6 (abstracted in *USSR Report: Space*, JPRS-USP-84-006, November 14, 1984, p. 7); "Salyut 7: Third Expedition to the Station," S.A. Bovin, *Zemlya i Vselennaya*, March-April 1985, pp. 9-15 (translated in *USSR Report*: Space, JPRS-USP-86-001, January 13, 1986, p. 58); *Zemlya i Vselennaya*, March-April 1985, pp. 15-22 (translated in *USSR Report*: Space, JPRS-USP-86-001, January 13, 1986, pp. 67, 69); "The Experience in Operation and Improving the Orlan-type Space Suits," I. P. Abramov, *Acta Astronautica*, Vol. 36, No. 1, July 1995, pp. 1-12.

August 30-September 5	**STS-41D/Discovery**
October 2	**Salyut 7/Soyuz-T 11 PE-3 landing**
October 5	**STS-41G/Challenger launch**

October 11
1984 EVA 12
World EVA 59
U.S. EVA 44
Shuttle EVA 6
Duration: 3:29
Spacecraft/mission: STS 41-G
Crew: Robert Crippen, Jon McBride, David Leestma, Kathryn Sullivan, Paul Scully-Power, Marc Garneau
Spacewalkers: David Leestma, Kathryn Sullivan
Purpose: Demonstrate the Orbital Refueling System for hydrazine fuel transfer

The astronauts reduced Challenger's cabin pressure to 70.3 kpascal (10.2 psi) 24 hr before planned airlock egress to reduce prebreathe time. Leestma and Sullivan entered the payload bay on mission day 7. The bay was largely taken up by the 10.7-m-by-2.1-m (35-ft-by-7-ft) Shuttle Imaging Radar (SIR)-B antenna, forcing both astronauts to attach their tethers to the port slidewire. This produced some minor tether crossing and tangling. The main task of the EVA was a test of the Orbiter Refueling System using toxic hydrazine fuel, which required 1 hr. The astronauts also tested the Provisional Stowage Assembly EVA tool box and new EMU boots. Before going inside they manually stowed the Ku-band antenna, which had given trouble earlier in the flight, and inspected the SIR-B antenna, which had not closed properly and had had to be pushed shut using the RMS. The astronauts found that insulation caught between the antenna sections was a possible cause. As they prepared to close out the EVA, the airlock hatch cover escaped. While Crippen maneuvered Challenger to pursue, Leestma somersaulted from the middle of the bay and snatched it. After airlock repressurization Leestma and Sullivan used the Interscan hydrazine vapor detector to confirm that they had carried no hydrazine inside on their EMUs.

STS 41-G Crew Report (no date); "Challenger Crew Obtains Significant Science Data," Craig Covault, *Aviation Week & Space Technology*, October 15, 1984, p. 16.

October 13	**STS-41G/Challenger landing**
November 8	**STS-51A/Discovery launch**

November 12
1984 EVA 13
World EVA 60
U.S. EVA 45

Shuttle EVA 7
MMU EVA 5
Duration: 6:00
Spacecraft/mission: STS 51-A
Crew: Frederick Hauck, David Walker, Joseph Allen, Anna Fisher, Dale Gardner
Spacewalkers: Joseph Allen, Dale Gardner
Purpose: Retrieve Palapa B-2 satellite

Before launch, Hauck, Discovery's commander for STS 51-A, gave the mission only a 50 percent chance of success because the satellites were not designed for retrieval nor EVA servicing. Only a month ahead of the mission, Flight Director Larry Bourgeois insisted that flight planners and crew develop plans for manually handling the satellite in case either of two handling aids, the Apogee Kick Motor Capture Device (the "stinger") and the "A-frame," failed to operate as planned. The STS 51-A crew and astronaut Jerry Ross developed a backup procedure and practiced it in the WETF and on JSC's air-bearing floor just three wk before flight. This EVA to recover the 555-kg (1222-lb) Palapa B-2 satellite took place on mission day 5 with Ross as EVA CapCom, Anna Fisher on the RMS, and David Walker as IV support crewman. Joseph Allen donned the MMU and attached the stinger to its arms. He then flew to Palapa, inserted the stinger into the spinning, drum-shaped satellite's Apogee Kick Motor bell, and activated the MMU's automatic attitude hold feature to stop the spin. Allen and Gardner cut off Palapa's omnidirectional antenna, then Gardner, standing in an MFR on the RMS, attempted to attach the 2.44-m (8-ft) A-frame device, which stubbornly resisted his efforts. The astronauts stowed the A-frame and fell back on the backup plan; Gardner grasped and held the satellite, then Fisher guided him to the stowage frame intended to hold Palapa in Discovery's payload bay. Gardner's gloves became abraded during the EVA because of friction from a knurled tool handle. The astronauts removed a bracket clamp from the A-frame during EVA closeout and took it into the airlock for examination in the crew compartment. Astronauts in the WETF began work to refine the backup satellite handling procedure for the Westar VI retrieval. Later investigation revealed that the A-frame was blocked by a Palapa waveguide extension that did not appear in the spacecraft blueprints.

STS 51-A Flightcrew Report (no date); "Satellite Retrieval Succeeds Despite Equipment Problem," *Aviation Week & Space Technology,* November 19, 1984, pp. 16-19; "The Fat Lady Sang," *Space News Roundup,* NASA JSC, December 7, 1984, p. 2; "Satellite Rescue Made Possible By Detailed Contingency Plans," Craig Covault, *Aviation Week & Space Technology,* December 10, 1984, pp. 46-49; "Shuttle 51A Missions Report," John Pfannerstill, *Spaceflight,* June 1985, pp. 260-265.

November 14

1984 EVA 14
World EVA 61
U.S. EVA 46
Shuttle EVA 8
MMU EVA 6
Duration: 5:42
Spacecraft/mission: STS 51-A
Crew: Frederick Hauck, David Walker, Joseph Allen, Anna Fisher, Dale Gardner
Spacewalkers: Joseph Allen, Dale Gardner
Purpose: Retrieve Westar VI satellite

This EVA to recover Westar VI marks the final flight of the MMU. By the time Gardner stowed the MMU, the two flight units had flown a total of 10 hrs, 22 min in space. Allen and Gardner ventured into Discovery's payload bay on mission day 7 with Anna Fisher again operating the RMS and Jerry Ross again serving as EVA CapCom. Gardner flew the MMU to the 499-kg

(1098-lb) Westar VI satellite and successfully stabilized it using the stinger and MMU. The astronauts left the omnidirectional antenna intact to serve as a handling aid. While Allen held the satellite on the RMS, Fisher moved him to the back of the payload bay where the astronauts secured Westar VI for return to Earth. A torque wrench escaped when Gardner bumped a tether release button, but IV crewman David Walker alerted him and he captured it before it could drift out of reach. In their postflight debrief, the crew noted that the EVAs were physically and mentally exhausting for all five STS 51-A crewmembers. The IV crewmembers were involved throughout the EVAs and had no opportunities to rest or eat, they reported. They suggested that Shuttle crews of at least six were desirable for missions with complex EVAs. Allen and Gardner reported that satellites were relatively easy for two crew to handle - in fact, they were easier to handle in some ways than the long, lightweight A-frame. Allen added that "As objects get smaller in space, they become more difficult to handle. It's really extraordinary how much easier it is to move more massive objects like satellites." The crew mission report states that

> [s]ince larger mass equates to greater stability, an object of light mass but large dimension - a very flimsy girder, for example - could prove the most challenging to handle since very gentle forces at one end could cause it to gyrate dramatically at the other end.

The crew also suggested that only contingency EVAs occur on mission day 1 or 2; that EVAs scheduled for mission day 3 be simple and short; that complex EVAs be performed no earlier than mission day 4; and that mission day 5 is the best first day for complex EVAs. JSC Director Gerald Griffin said after the flight that the "flexibility that people bring to the equation has made. . . the ultimate difference. We have proven that the flexibility of people in orbit allows us to respond in short order to unforeseen circumstances." He added that the Solar Max and Palapa/Westar flights "taught us a great deal about what we can do with people in low earth orbit, and this knowledge will prove useful to us as we construct the space station."

STS 51-A Flightcrew Report (no date); *Astronautics and Aeronautics, 1979-1984*, NASA, 1990, p. 516; "Satellite Retrieval Succeeds Despite Equipment Problem," *Aviation Week & Space Technology*, November 19, 1984, pp. 16-19; "The Fat Lady Sang," *Space News Roundup*, NASA JSC, December 7, 1984, p. 2; "Satellite Rescue Made Possible By Detailed Contingency Plans," Craig Covault, *Aviation Week & Space Technology*, December 10, 1984, pp. 46-49.

November 16	**STS-51A/Discovery landing**

1985

January 24-27	**STS-51C/Discovery**
April 12	**STS-51D/Discovery launch**
April 16	

1985 EVA 1
World EVA 62
U.S. EVA 47
Shuttle EVA 9
Duration: 3:06
Spacecraft/mission: STS 51-D
Crew: Karol Bobko, Donald Williams, Jeffrey Hoffman, David Griggs, Rhea Seddon, Charles Walker, Jake Garn
Spacewalkers: Jeffrey Hoffman, David Griggs

Purpose: Perform first unscheduled EVA; install improvised switch-pulling appendages on RMS

One of Discovery's two satellite payloads on mission STS 51-D, the Hughes Syncom-IV/Leasat 3 geosynchronous communications satellite, remained inert after deployment on April 13 and engineers quickly determined that a likely culprit was a faulty switch on the satellite's side. The spate of malfunctions in Shuttle-launched satellites encouraged managers to authorize the first unscheduled EVA of the Shuttle program. Engineers first proposed that an astronaut ride the RMS close to the satellite and trip the switch by hand using an improvised tool. Discovery carried no MFR, however, and astronauts Sherwood Spring and Jerry Ross, working in the WETF, were unable to develop a satisfactory substitute based on materials aboard Discovery. Engineers then hatched a plan in which the EVA astronauts would attach improvised switch-pulling appendages to the RMS end effector. Rhea Seddon would maneuver the RMS so that the appendages snag the switch that started an automatic timer, and Discovery would quickly move away to get clear of the plume from the satellite's kick motor. Griggs and Hoffman trained for contingency EVAs on the canceled STS 41-F and 51-E missions before they were assigned to STS 51-D, so they each had over 50 hr of contingency EVA training - more than four times as much as normal - going into this EVA. EVA operations and in-flight maintenance engineers developed and designed procedures. Ross and Spring simulated the procedure in the WETF, then talked through the planned EVA procedure with Griggs and Hoffman over the radio. The astronauts discussed the procedures between themselves. Two different switch-pullers (the "lacrosse stick" and the "flyswatter") were manufactured using materials on board Discovery, including tape and plastic book covers. Twenty-four hr ahead of the EVA the crew reduced Discovery's cabin pressure to 70.3 kpascal (10.2 psi) to cut prebreathe time to 1 hr. Hoffman called prebreathing "a miserable experience." The thrill of an unexpected trip into Discovery's payload bay quickly overshadowed his discomfort, however. According to his taped diary:

> Going out of the airlock. . . was very much like the water tank because we tended to be looking down, not so much aware of the world going by. It was when I went up to the sill [where the payload bay doors are attached] to start moving back to the work position that I really turned myself upside-down and looked at the Earth going by. That's when the true awesomeness of what we were doing really struck me. That and having an almost 180-degree panorama of the entire world, our spaceship, and me just floating there. . . there was a part of me just standing back not believing what I was seeing and what I was doing.

Before the EVA Seddon bent the RMS at elbow and wrist to bring the RMS end effector closer to the airlock. The astronauts moved to the end effector, which is about halfway down the payload bay on the starboard side, and attached the switch-pullers with a payload retention strap while EVA CapCom Jerry Ross advised. Griggs and Hoffman rested while they were out of communication range. During night passes both men became cold, as Hoffman describes in his diary:

> . . . Sitting around at night I turned the cooling down all the way to keep it as warm as possible. My body was quite comfortable but the tips of my fingers got quite cold. Dave's apparently got even colder.

Their work completed, Griggs pushed himself over the starboard payload bay sill toward the orbiter's delicate radiator panels, and Commander Karol Bobko warned him to return to the payload bay. The astronauts then returned to the warmth of Discovery's crew compartment. Rhea Seddon snared the switch repeatedly with the switch-pullers, but Syncom IV/Leasat 3 remained inert.

STS 51-D Flightcrew Report (no date); *An Astronaut's Diary*, Jeffrey Hoffman, 1986, pp. 30-33; "Shuttle 51D Mission Report," John Pfannerstill, *Spaceflight*, November 1985, pp. 414-419; email from Charles Walker, September 10, 1996; interview, David S. F. Portree with Jeff Hoffman, June 18, 1996.

April 19	STS-51D/Discovery landing
April 29-May 6	STS-51B/Challenger
June 6	Salyut 7/Soyuz-T 13 PE-4 launch
June 17-24	STS-51G/Discovery
July 29-August 6	STS-51F/Challenger

August 2
1985 EVA 2
World EVA 63
Russian EVA 16
Space Station EVA 24
Duration: 5:00
Spacecraft/mission: Salyut 7 PE-4
Crew: Vladimir Dzhanibekov, Viktor Savinykh
Spacewalkers: Vladimir Dzhanibekov, Viktor Savinykh
Purpose: Augment Salyut 7 solar arrays; test Orlan-DM suits

This EVA marked the first use of the Orlan-DM space suit designed for deployment on the Mir space station. Orlan-DM featured many improvements over the Orlan-D, including bright lights at the temples of the "headset" for illuminating suit control dials; improved controls; sturdier construction, including rubberized fabric shoulder belts in place of the Orlan-D's rubber belts; and greater mobility. The suits reached Salyut 7 aboard Cosmos 1669 (July 21, 1985), a prototype Progress freighter improved for Mir. During this EVA, Savinykh and Dzhanibekov augmented the port side solar array using two extension panels delivered by Progress 24 (docked June 23, 1985). One extension had an experimental design. This completed the series of solar array augmentation spacewalks planned at Salyut 7's launch to occur over the station's occupancy. Moscow TV showed portions of the EVA live. The cosmonauts used the Orlan-DM's headset lights to continue work during orbital night. They left a small piece of solar cell material outside as an exposure experiment. Before closing out the EVA, they installed a Soviet/French experiment for collecting meteoritic dust (it was expected to gather dust from Halley's Comet) and changed space exposure cassettes near the transfer compartment hatch.

Izvestiya, August 4, 1985, p. 1 (translated in *USSR Report: Space*, JPRS-USP-86-001, January 13, 1986, pp. 16-17); "Salyut Mission Report," *Spaceflight*, Neville Kidger, December 1985; *Pravda*, August 3, 1985, p. 3 (translated in *USSR Report: Space*, JPRS-USP-86-001, January 13, 1986, p. 18); *Izvestiya*, August 4, 1985, p. 2 (translated in *USSR Report: Space*, JPRS-USP-86-001, January 13, 1986, p. 2).

| August 27 | STS-51I/Discovery launch |

August 31
1985 EVA 3
World EVA 64
U.S. EVA 48
Shuttle EVA 10

Duration: 7:20
Spacecraft/mission: STS 51-I
Crew: Joe Engle, Richard Covey, James van Hoften, William Fisher, John Lounge
Spacewalkers: William Fisher, James van Hoften
Purpose: Retrieve Leasat 3 satellite; begin repairs

Failure of Syncom-IV/Leasat 3 on STS 51-D in April was followed by four months of intense preparation for a repair EVA. Only a single EVA was planned, but shortly after Discovery reached orbit on August 27 RMS operator Mike Lounge discovered that fuses had blown in the arm, forcing him to position its joints one at a time with no computer assistance. On August 28 mission controllers in Houston determined that two EVAs were necessary to complete the repair and began replanning the mission. Van Hoften and Fisher checked out their suits on August 30. On this date, Discovery made rendezvous with the satellite in a 400-by-290-km (250-by-180-mi) orbit. The astronauts entered Discovery's payload bay and Van Hoften placed his feet in the MFR. The 6818-kg (15,000-lb) drum-shaped satellite rotated very slowly, so Van Hoften found it easy to install a bar for slowing rotation by hand. However, the EVA fell behind schedule because the handling bar - which included a grapple fixture for the RMS end-effector - did not fit at first. The crew complained of being too cold. Van Hoften and Fisher safed the satellite using plugs and specialized tools, then installed a bypass cable harness to work around the faulty switches that prevented activation in April. They discovered that the satellite's batteries had not frozen as some had feared. Syncom-IV/Leasat 3's omnidirectional antenna popped up, indicating a successful repair, and Van Hoften and Fisher closed out the first EVA. The crew left Leasat 3 safed on the RMS when they bedded down for the night.

Ox's Crew Report (no date); *Flight 51-I Mission Report of Leasat EVA Activities*, Crew Systems Division, October 8, 1985; "Shuttle 51-I Mission Report," Roelof Schuiling, *Spaceflight*, December 1985, pp. 466-468; "Astronauts Repair, Deploy Leasat During Two Space Shuttle EVAs," Craig Covault, *Aviation Week & Space Technology*, September 9, 1985, pp. 21-23.

September 1
1985 EVA 4
World EVA 65
U.S. EVA 49
Shuttle EVA 11
Duration: 4:26
Spacecraft/mission: STS 51-I
Crew: Joe Engle, Richard Covey, James van Hoften, William Fisher, John Lounge
Spacewalkers: William Fisher, James van Hoften
Purpose: Complete repair of Leasat 3 satellite, release

Van Hoften and Fisher installed an instrumented cover over Syncom-IV/Leasat 3's apogee kick motor nozzle, then armed the motor. The astronauts experienced difficulties handling the satellite, which threatened to collide with Discovery. This was largely because they could not see each other from their positions on opposite sides of the 4.3 m-dia (14 ft-dia) satellite and thus imparted opposing motions. Van Hoften warned that "if something happens and I'm about to lose it, I'm going to give it a heck of a push and bail out." The astronauts managed to control the satellite's motions, however. Van Hoften spun up Syncom-IV/Leasat 3 manually to 3 rpm and released it. In their postflight debrief, the astronauts recommended against EVAs on consecutive days, and stated that the EMU is "unquestionably overcooled." In his crew report, Van Hoften stated that his fingers became very cold while he held Leasat 3, even with water to his LCVG shut off. With the water off, his EMU helmet fogged up. Syncom-IV/Leasat 3 proceeded successfully to geosynchronous orbit after warming up in low-Earth orbit for several months.

Ox's Crew Report (no date); "STS 51-I Mission Report," Roelof Schuiling, *Spaceflight*, December 1985, p. 468; "Astronauts Repair, Deploy Leasat During Two Space Shuttle EVAs," Craig Covault, *Aviation Week & Space Technology*, September 9, 1985, pp. 21-23.

September 3	**STS-51I/Discovery landing**
September 17	**Salyut 7/Soyuz-T 14 PE-5 launch**
September 26	**Salyut 7/Soyuz-T 13 PE-4 landing**
October 3-7	**STS-51J/Atlantis**
October 30-November 6	**STS-61A/Challenger**
November 21	**Salyut 7/Soyuz-T 14 PE-5 landing**
November 26	**STS-61B/Atlantis launch**

November 29

1985 EVA 5
World EVA 66
U.S. EVA 50
Shuttle EVA 12
Duration: 5:32
Spacecraft/mission: STS 61-B
Crew: Brewster Shaw, Bryan O'Connor, Sherwood Spring, Mary Cleave, Jerry Ross, Charles Walker, Rodolfo Neri Vela
Spacewalkers: Jerry Ross, Sherwood Spring
Purpose: Assemble experimental erectable truss structures

The STS 61-B EVAs were designed to demonstrate assembly techniques which might be used in space station assembly. This first EVA focused on human performance. The assembly procedures were precisely timed ahead of the EVAs in NASA neutral buoyancy facilities to determine if underwater simulation verifies EVA performance. In addition, the EMUs were instrumented to allow precise monitoring of oxygen consumption during work. The astronauts first assembled the 3.4-m (11-ft) Assembly Concept for Construction of Erectable Space Structures (ACCESS) assembly jig in Atlantis' payload bay. Each cell was assembled in the jig, then pushed up so that the next cell could be assembled. Ross later called this "a neat way to build a truss." Assembling the ACCESS truss required 58 min in the water tank, and the EVA timeline allotted 2 hr for a single ACCESS assembly. Only 55 min were required to build the truss, however, so the astronauts disassembled it and built it again. The Experimental Assembly of Structures through EVA (EASE) task assessed the capabilities of free-floating astronauts, and involved putting together beams weighing 29 kg (64 lb) to make a 3.6-m (12-ft) three-sided pyramid. EASE was scheduled to be assembled six times, but the astronauts managed eight assemblies. During the first four assemblies the astronauts used foot restraints. Spring noted in his postflight debriefing that his fingers grew numb during the third EASE assembly and very tired during the fourth. At the end of the EVA Spring assembled and hand-deployed a small target satellite to be used after the EVA as a station-keeping target for Atlantis, which played the role of an automated orbital maneuvering vehicle in rendezvous software tests.

EASE/ACCESS Postmission Management Report, NASA Marshall Space Flight Center (MSFC) (no date); "Shuttle Mission EVAs to Demonstrate Space Station Assembly Techniques," Craig Covault, *Aviation Week*

& *Space Technology*, November 25, 1985, pp. 63-69; "Shuttle EVAs Utilize Techniques Planned for Space Station Assembly," Craig Covault, *Aviation Week & Space Technology*, December 9, 1985, pp. 21-23; interview David S. F. Portree with Jerry Ross, January 11, 1996.

December 1

1985 EVA 6
World EVA 67
U.S. EVA 51
Shuttle EVA 13
Duration: 6:41
Spacecraft/mission: STS 61-B
Crew: Brewster Shaw, Bryan O'Connor, Sherwood Spring, Mary Cleave, Jerry Ross, Charles Walker, Rodolfo Neri Vela
Spacewalkers: Jerry Ross, Sherwood Spring
Purpose: Assemble and manipulate experimental erectable truss structures; demonstrate RMS use in assembly

The second EASE/ACCESS EVA sought to assess the ability of astronauts to handle large structural elements and the ability of the RMS to support station assembly. Ross and Spring assembled nine bays of ACCESS, then placed parts for the tenth bay on the RMS. Ross stepped into the MFR and Mary Cleave positioned him within reach of the top of the ACCESS girder, where he assembled the tenth bay. The parts were not tethered. Ross performed a cable run assembly simulation by attaching a tether along the side of the tower while Cleave positioned him. Then Spring released the bottom of the tower so Ross could try to precisely handle the beam from the RMS. He replaced it in the assembly jig where it started, demonstrating astronaut ability to assemble a truss in one place and install it in another. Spring then replaced Ross on the MFR. He changed a beam on the tower to simulate structural repair, then pointed the truss at the Moon to judge his handling ability. The astronauts took down ACCESS, and Spring assembled EASE from the RMS. Before finishing, he joined two beams to simulate handling a thermal control heat pipe. Ross unlatched the EASE pyramid so that his partner could maneuver it. Then he replaced Spring on the MFR to duplicate the EASE activities. "This is probably not the preferred way of building a space station," Ross said later of EASE. The astronauts reported that the most difficult part of the EVAs was torquing their own 182-kg (400-lb) masses while holding the EASE beams. Generally speaking, ACCESS worked well, while EASE required too much freefloating. The astronauts judged that performing 6-hr EVAs every other day over a 5- or 6-day period was feasible, and recommended glove changes to reduce hand fatigue. Ross said in the EVA debrief that the crew had tried to have the MMU manifested for use in the second EVA, because "for certain applications it would be very useful. . . in particular if you were building portions of a space station attached to the orbiter, then moving those portions farther than the manipulator arm could transport them." He added that the MMU could be used to attach cable runs and instruments in places out of reach of the RMS.

EASE/ACCESS Postmission Management Report, NASA MSFC (no date); "Shuttle Mission EVAs to Demonstrate Space Station Assembly Techniques," Craig Covault, *Aviation Week & Space Technology*, November 25, 1985, pp. 63-69; "Shuttle EVAs Utilize Techniques Planned for Space Station Assembly," Craig Covault, *Aviation Week & Space Technology*, December 9, 1985, pp. 21-23; "Astronauts Believe Lengthy EVA Building Sessions are Feasible," *Aviation Week & Space Technology*, December 16, 1985, p. 20-21; interview, David S. F. Portree with Jerry Ross, January 11, 1996.

December 3   STS-61B/Atlantis landing

1986

January 12-18	**STS-61C/Columbia**
January 28	**STS-51L/Challenger**
March 13	**Mir/Soyuz-T 15 PE-1 launch**
May 5-6	**Soyuz-T 15 transfer from Mir to Salyut 7**

May 28
1986 EVA 1
World EVA 68
Russian EVA 17
Space Station EVA 25
Duration: 3:50
Spacecraft/mission: Salyut 7 PE-6
Crew: Leonid Kizim, Vladimir Solovyov
Spacewalkers: Leonid Kizim, Vladimir Solovyov
Purpose: Test experimental deployable truss; remove space exposure cassettes from Salyut 7's exterior for return to Earth

Seven-time EVA veterans Solovyov and Kizim first visited the new Mir station then transferred to Salyut 7 to tie up loose ends left by the station's previous crew, which had been forced to end its mission early after its commander, Vladimir Vasyutin, became ill. Vasyutin and Alexandr Volkov were to have performed EVA assembly experiments outside Salyut 7 while Viktor Savinykh assisted from inside Salyut 7. Solovyov and Kizim removed and placed inside the transfer compartment space exposure cassettes and the joint Soviet-French micrometeoroid collector deployed in August 1985. The exposure cassettes included

- Spiral, for study of space effects on cables

- Istok, which looked for changes in threaded connectors (nuts and bolts) caused by space exposure

- Resurs, which studied space effects on metals

- Meduza, which studied space effects on biopolymers

The cosmonauts then attached the cylindrical 150-kg (330-lb) URS space assembly device to the hull outside the airlock hatch. The URS device deployed a 20-kg (44-lb), 12-to-15-m (40-to-50-ft) tubular metal truss held together by hinges and springs. URS was designed and built by the Paton Institute of Electric Welding in Kiev, which also developed the URI tool used by Savitskaya during her 1984 EVA. The URS truss was deployable, as opposed to the erectable EASE and ACCESS systems Ross and Spring worked with 7 mo earlier on STS 61-B. *Pravda*, quoting Paton Institute sources, stated that the truss' length could be increased to a kilometer or more by adding more folded cassettes. Kizim operated the three buttons that controlled deployment, then climbed halfway up the truss. He found it sturdy, with oscillations limited to a few centimeters of amplitude. The top of the URS truss carried Leningrad Polytechnical Institute's 3-kg (6.6-lb) Fon ("Background") device, which assessed the environment around Salyut 7. The cosmonauts installed the BOSS visible light communications system on a work compartment porthole, then refolded the URS girder and closed out the EVA. Portions of the EVA were televised live to Soviet audiences. They spent the next two days cleaning their Orlan-DM suits, undergoing debriefing, and preparing documentation.

"First Soviet Structure in Space," *Air & Cosmos*, June 28, 1986, p. 52 (translated from French in *USSR Report: Space*, JPRS-USP-86-007-L, October 7, 1986, pp. 1-2); "Cosmonauts Deploy Girder from Salyut 7," *Pravda*, May 29, 1986, p. 1 (translated in *USSR Report: Space*, JPRS-USP-86-006, November 12, 1986, p. 1); "Developer Comments on Girder Deployment Experiment," A. Tarasov, Pravda, May 29, 1986, pp. 1, 6 (translated in *USSR Report: Space*, JPRS-USP-86-006, November 12, 1986, p. 2); "New Structure for Mir," Neville Kidger, *Spaceflight*, September/October 1986, pp. 346-347; "Problems in the Exploitation of Space," Sergei Grishin and Sergei Chekalin, *Novoye v Zhizni, Nauke, Tekhnike: Seriya Kosmonavtika, Astronomiya*, January 1988 (excerpted and translated in *JPRS Report, Science & Technology, USSR: Space*, August 17, 1988, pp. 40-43).

May 31
1986 EVA 2
World EVA 69
Russian EVA 18
Space Station EVA 26
Duration: 4:40
Spacecraft/mission: Salyut 7 PE-6
Crew: Leonid Kizim, Vladimir Solovyov
Spacewalkers: Leonid Kizim, Vladimir Solovyov
Purpose: Test experimental deployable truss; test URI tool

This eighteenth Soviet EVA was the ninth for the Kizim-Solovyov team and the last carried out on Salyut 7. The cosmonauts extended the URS truss, then used the BOSS device installed on the previous EVA to relay data on truss stability from instruments at the top of the truss. These included a small seismograph built by the All-Union Scientific Research Institute of Geophysics for tracking low-frequency (small) vibrations imparted by the station's acceleration, and the Mayak ("beacon") experiment, in which a camera tracked high-frequency vibrations by filming the movements of a small orange light attached to the truss. Solovyov and Kizim then rigidized the truss by welding portions using the URI tool. After closing and dismantling the truss, they installed the Mikrodeformator device built by Kharkov Polytechnical Institute, which studied aluminum-magnesium alloy reactions to repeated ("complex cyclical") structural loads under space conditions. At the end of the EVA, they brought inside the sample of solar cell material left outside by Savinykh and Dzhanibekov in August 1985. According to *Pravda*, ". . .successful accomplishment of multifaceted experimental operations in open space confirms the prospects of the technological operations that have been developed, as well as the possibility of their practical application in creating complex, large-size orbiting complexes for scientific and economic purposes." The Paton Institute's V. Lapchinsky asserted that, "[w]e are at the threshold of the era of space construction."

"Cosmonauts Continue Girder Experiments in Second EVA," *Pravda*, June 1, 1986, p. 1, 4 (translated in *USSR Report: Space*, JPRS-USP-86-006, November 12, 1986, p. 6); "Commentary on Experiments in Second Girder Deployment," *Pravda*, June 1, 1986, p. 6 (translated in *USSR Report: Space*, JPRS-USP-86-006, November 12, 1986, p. 6); "Star Construction Project: Salyut 7, Mir - Our Commentary," B. Paton and Yu. Semenov, *Pravda*, August 16, 1986 (translated in *USSR Report: Space*, JPRS-USP-86-006, November 12, 1986, p. 33); "New Structure for Mir," Neville Kidger, *Spaceflight*, September/October 1986, pp. 346-347; "The Experience in Operation and Improving the Orlan-type Space Suits," I. P. Abramov, *Acta Astronautica*, Vol. 36, No. 1, July 1995, pp. 1-12.

June 25-26   **Soyuz-T 15 transfer from Salyut 7 to Mir**

1987

February 5   Mir/Soyuz-TM 2 PE-2 launch

April 11
1987 EVA 1
World EVA 70
Russian EVA 19
Space Station EVA 27
Duration: 3:35
Spacecraft/mission: Mir PE-2
Crew: Yuri Romanenko, Alexander Laveikin
Spacewalkers: Yuri Romanenko, Alexander Laveikin
Purpose: Contingency EVA to investigate cause of Kvant module hard docking difficulties

The first Mir space station EVA - only the 19th of the Soviet space program - was a contingency spacewalk to permit Kvant, the station's first expansion module, to complete docking. One of its participants, Yuri Romanenko, took part in the first Soviet space station EVA aboard Salyut 6 in 1977, which also involved a docking unit inspection. Kvant arrived at the Mir core module's aft port early on April 10 attached to a Functional Service Module (FSM) space tug. It achieved soft dock, but full retraction of the Kvant probe proved impossible, and the docking collars remained separated by a few centimeters. The cosmonauts observed Kvant from Mir's aft compartment viewports, but were unable to detect anything out of the ordinary. The Kvant-FSM combination was left with its probe latched in the aft port's Konus drogue unit while controllers in the TsUP investigated the problem. Attitude control maneuvers were suspended because Kvant might pivot in the Konus drogue, banging together the docking collars. A contingency EVA was quickly authorized, and the cosmonauts spent April 11 making preparations. Just before midnight Moscow time on this date, Romanenko and Laveikin left one of the four berthing ports in the forward transfer compartment and moved 13 m (43 ft) along Mir's hull to the aft port. Cosmonauts in the Hydrolaboratory in Star City carried out the procedure simultaneously. Flight Engineer Laveikin's Orlan-DM space suit registered a minor pressure drop, causing momentary concern, but the problem was quickly traced to an incorrect switch setting. Flight controllers extended the docking probe to permit the cosmonauts to examine the docking unit. They discovered an "extraneous white object" jammed between the two spacecraft. This was later identified as a twisted piece of cloth, possibly trash escaped from Progress 28, which had undocked from the aft port on March 26. With difficulty Laveikin freed and discarded the object early in the morning on April 12, the Cosmonautics Day holiday in the Soviet Union. The cosmonauts waited nearby while the TsUP commanded the Kvant probe to retract, completing hard dock, then returned inside an expanded Mir station.

The Soviet Year in Space 1987, Nicholas Johnson, 1988, pp. 88-89; "Spacewalk Saves Mission," Neville Kidger, *Spaceflight*, June 1987, pp. 236-237; "Mir Mission: Third Solar Array Installed," *Spaceflight*, August 1987, pp. 284.

June 12
1987 EVA 2
World EVA 71
Russian EVA 20

Space Station EVA 28
Duration: 1:53
Spacecraft/mission: Mir PE-2
Crew: Yuri Romanenko, Alexander Laveikin
Spacewalkers: Yuri Romanenko, Alexander Laveikin
Purpose: Install Mir top solar array

To save weight, the Mir base block was launched with only two solar arrays. These provided a total of only 9.4kW of electricity, leaving the base block hungry for power. The base block had a socket on top for a third array delivered inside the Kvant module. On June 9 Romanenko and Laveikin capped a busy day spent studying Supernova 1987A using the Kvant module's Roentgen observatory by undergoing a medical checkup to certify them fit for two solar array installation EVAs. Spacewalk preparations alternated with astronomical observations on June 10 and 11, and lasted all day on this date. The first installation EVA began late in the evening. Romanenko and Laveikin attached an extendible "hinged lattice girder" truss to the top of the Mir complex, then attached folded solar panels to both sides of the girder. To test their ability to operate without EVA restraints, Romanenko and Laveikin employed no foot restraints on this and their next EVA, relying instead on tethers. Laveikin later stated that this gave them "more freedom to maneuver, but we had to cling to the ship with one hand."

Izvestiya, June 13, 1987, p. 3 (translated in *JPRS Report, Science & Technology, USSR: Space*, JPRS-USP-87-005, August 19, 1987, p. 12); *Izvestiya*, June 10, 1987, p. 1 (translated in *JPRS Report, Science & Technology, USSR: Space*, JPRS-USP-87-005, August 19, 1987, p. 10); *Sovetskaya Rossiya*, June 13, 1987, p. 1 (translated in *JPRS Report, Science & Technology, USSR: Space*, JPRS-USP-87-005, August 19, 1987, p. 11); *Izvestiya*, June 14, 1987, p. 1 (translated in *JPRS Report, Science & Technology, USSR: Space*, JPRS-USP-87-005, August 19, 1987, p. 13); *Kosomolskaya Pravda*, S. Leskov, February 27, 1988, p. 1 (translated in *JPRS Report, Science & Technology, USSR: Space*, August 17, 1988, p. 6).

June 16
1987 EVA 3
World EVA 72
Russian EVA 21
Space Station EVA 29
Duration: 3:15
Spacecraft/mission: Mir PE-2
Crew: Yuri Romanenko, Alexander Laveikin
Spacewalkers: Yuri Romanenko, Alexander Laveikin
Purpose: Complete installation of Mir top solar array

Romanenko and Laveikin placed an extendible truss on top of the one they installed on June 12 and attached folded solar arrays to either side. They linked the electrical systems of the array sections, then deployed the structure to its full height of 10.6 m (35 ft) "using special mechanisms." Each of the four array sections was made up of eight rectangular solar cell leaves with a total area of about 24 sq m (258 sq ft). Before going inside, Romanenko and Laveikin attached space exposure cassettes to Mir's exterior. By June 23 the cosmonauts completed work inside Mir to connect the new array to Mir's electrical system. The completed array increased available power by 2.4kW. According to *Pravda*, "[i]ncreasing the capacity of the onboard power supply system helps to increase the effectiveness of the scientific research work on the Mir complex substantially."

Gudok, B. Kutznetsov, June 16, 1987, p. 4 (translated in *JPRS Report, Science & Technology, USSR: Space*, JPRS-USP-87-005, August 19, 1987, p. 16); *Pravda*, June 18, 1987, p. 1 (translated in *JPRS Report,*

Science & Technology, USSR: Space, JPRS-USP-87-005, August 19, 1987, p. 14-15); *Pravda*, June 24, 1987, p. 1 (translated in *JPRS Report, Science & Technology, USSR: Space*, JPRS-USP-87-005, August 19, 1987, p. 18).

July 22	**Mir/Soyuz-TM 3 VE-1 launch**
July 30	**Mir/Soyuz-TM 2 VE-1 landing**
December 21	**Mir/Soyuz-TM 4 PE-3 launch**
December 29	**Mir/Soyuz-TM 3 PE-2 landing**

1988

February 26
1988 EVA 1
World EVA 73
Russian EVA 22
Space Station EVA 30
Duration: 4:25
Spacecraft/mission: Mir PE-3
Crew: Vladimir Titov, Musa Manarov
Spacewalkers: Vladimir Titov, Musa Manarov
Purpose: Replace section of Mir solar array; inspect Mir's exterior

On February 15 Titov and Manarov underwent an inflight refresher course in the fine art of changing solar array sections by watching a videotape of their own preflight practice sessions in the Hydrolaboratory. On February 19 and 23 they inspected their Orlan-DM space suits. On this date they opened one of the four radial berthing ports in Mir's transfer compartment while out of communication with the TsUP, prepared their work site at the base of the solar array installed by the Mir PE-2 cosmonauts in June, and replaced one of four sections of the array. This entailed "collapsing" the lower extendible boom to fold shut both solar array sections attached to it. The new section was, like the one it replaced, made up of eight leaves of solar cells. Carbon-plastic composite replaced metal in the new section, however, and six of the leaves used improved solar cells that produced as much power as eight conventional leaves while better withstanding the rigors of space. The remaining two leaves were instrumented and independently replaceable, providing a test site for new solar cell materials. The cosmonauts stood in foot restraints while they worked, continuing the EVA restraint tests begun on Mir PE-2. They redeployed the extendible boom, unfolding the new section and exposing it to sunlight. To round out the EVA, Manarov and Titov moved back along the Kvant module to inspect the rendezvous antenna on Progress 34 (it was late in opening), televised Mir's exterior and the Soyuz TM-4 spacecraft for the benefit of engineers on Earth, and replaced space exposure cassettes.

"Bulgarian Set for Mir Visit," Neville Kidger, *Spaceflight*, June 1988, pp. 228-229; *Sotsialisticeskaya Industriya*, February 13, 1988, p. 4 (translated in *JPRS Report, Science & Technology, USSR: Space*, August 17, 1988, pp. 4-5); *Izvestiya*, A. Ivakhnov, February 28, 1988, p. 1 (translated in *JPRS Report, Science & Technology, USSR: Space*, JPRS-USP-88-003, August 18, 1988, p. 6); *Trud*, V. Golovachev, February 27, 1988, p. 1 (abstracted in *JPRS Report, Science & Technology, USSR: Space*, JPRS-USP-88-003, August 17, 1988, pp. 6-7); *Kosomolskaya Pravda*, S. Leskov, February 27, 1988, p. 1 (translated in *JPRS Report, Science & Technology, USSR: Space*, August 17, 1988, p. 6); *Pravda*, February 27, 1988, p. 1 (translated in *JPRS Report, Science & Technology, USSR: Space*, August 17, 1988, p. 7); *Izvestiya*,

February 24, 1988, p. 1 (translated in *JPRS Report, Science & Technology, USSR: Space*, August 17, 1988, p. 7); Pravda, February 20, 1988, p. 1 (translated in *JPRS Report, Science & Technology, USSR: Space*, August 17, 1988, p. 7).

June 7	**Mir/Soyuz-TM 5 VE-2 launch**
June 17	**Mir/Soyuz-TM 4 VE-2 landing**

June 30
1988 EVA 2
World EVA 74
Russian EVA 23
Space Station EVA 31
Duration: 5:10
Spacecraft/mission: Mir PE-3
Crew: Vladimir Titov, Musa Manarov
Spacewalkers: Vladimir Titov, Musa Manarov
Purpose: Repair TTM X-ray detector on Kvant

The joint Dutch-British-Soviet TTM X-ray telescope gave trouble soon after launch on the Kvant module in April 1987, so engineers proposed and received approval for an EVA to replace its detector. The TTM telescope was not designed for EVA servicing. Some tools for the repair were developed by Dutch and Soviet scientists and delivered by the Mir VE-2 crew. Before going outside, Titov and Manarov received a familiarization briefing from British researchers who helped design and build the detector. During the EVA Dutch TTM researchers stood by in the TsUP. The cosmonauts cut through 20 layers of thermal insulation to reach the 40-kg (88-lb) detector. Because there were no footholds or handholds at the worksite, they took turns working while the other held him. More clips held the detector in place than expected. Three screws locked in place by resin threw them off timeline; they had to scrape one with a saw blade before it would turn, and the effort required to turn the screws forced them to rest several times. After the cosmonauts accomplished 70 percent of the task a special "key" tool for removing a brass clamp snapped. Before they passed out of radio contact, the TsUP gave the cosmonauts 15 min to remove the clamp using other tools. When communication was restored, Titov and Manarov reported that they had given up and returned to the transfer compartment hatch. Before entering the airlock, they measured attachment locations for a foot restraint to be used on an upcoming Soviet-French spacewalk. Two French specialists monitored this part of the EVA in the TsUP. Titov's Orlan-DM suit gave him a false "ventilation low" signal caused when humidity interfered with a sensor. This EVA marked the last use of the Orlan-DM space suit.

"Mir Mission Report: TTM Telescope Repaired," Neville Kidger. *Spaceflight*, February 1989, p. 64; "Soviet Cosmonauts on Mir Fail to Repair Science Instrument," *Aviation Week & Space Technology*, July 11, 1988, p. 27; *The Soviet Year in Space 1988*, Nicholas Johnson, 1989, p. 95; "The Experience in Operation and Improving the Orlan-type Space Suits," I. P. Abramov, *Acta Astronautica*, Vol. 36, No. 1, July 1995, pp. 1-12.

August 29	**Mir/Soyuz-TM 6 VE-3 launch**
September 7	**Mir/Soyuz-TM 5 VE-3 landing**
September 29-October 3	**STS-26/Discovery**

October 20

1988 EVA 3
World EVA 75
Russian EVA 24
Space Station EVA 32
Duration: 4:12
Spacecraft/mission: Mir PE-3
Crew: Vladimir Titov, Musa Manarov, Valeri Polyakov
Spacewalkers: Vladimir Titov, Musa Manarov
Purpose: Repair TTM X-ray telescope; test Orlan-DMA space suit

A second TTM repair EVA was originally set for July 5, but was postponed to permit more preparation. On September 9, Progress 38 delivered seven new tools and the first Orlan-DMA space suits. Orlan-DMA was an upgrade of the short-lived model Orlan-DM (1985-1988), which was itself an upgrade of Orlan-D (1977-1985). Like earlier Orlan models, Orlan-DMA retained the distinctive rear-entry hatch built into its hard aluminum alloy torso. A cable lanyard and locking handle were used to close and seal the rear hatch. Orlan-DMA's life support system activated when the handle locked into place. Improvements included:

- Composite fabric in the arms and legs was lighter, more flexible, and tougher than previously used fabrics. Arms and legs could be removed for repair or replacement. The suit was sized for specific cosmonauts by pulling or releasing cables and pulleys in the arms and legs.

- In the event of glove puncture, a forearm cuff inflated around the wrist using air from the backup oxygen tank, sealing off the cosmonaut's glove until he could return to the airlock; though painful, this was certified by volunteers in a vacuum chamber as a life-saving system.

- More durable life support system electrical motors.

- Improved gloves for better hand mobility. Gloves were custom-made for each cosmonaut (some sources, however, state that only two sizes were available).

The Orlan-DMA weighed 105 kg (231 lb) fully charged and 90 kg (198 lb) empty. The integral backpack measured 1.19 m (3.9 ft) long and 48 cm (18.9 in) wide. The suit had a maximum operating pressure of 40 kilopascal (5.8 psi) and a minimum pressure of 26.2 kilopascal (3.8 psi). Typical EVA duration was 6 to 7 hr, up from 5 hr for the Orlan-DM. Like the Orlan-D and Orlan-DM suits before it, Orlan-DMA had dual polyurethane rubber pressure bladders, one inside the other. The inner bladder inflated only if the primary layer was punctured. A replaceable lithium hydroxide cartridge absorbed exhaled carbon dioxide. Like earlier Orlan models, Orlan-DMA's liquid-cooling garment coverall had an integral head covering. Voice communication was by the Korona system, which included two microphones, two earphones, and primary and backup transceivers and amplifiers. Korona's antenna was embedded in the suit's outer fabric layer. Orlan-DMA's chief improvement was its add-on radio and battery package for making the suit autonomous. Both Orlan-D and Orlan-DM relied on an umbilical connection with the space station for their electricity and communications and to supply the ground with telemetry on cosmonaut and suit health. The add-on package was phased in during 1990 so that Orlan-DMA could be used with the SPK maneuvering unit, the Soviet equivalent of the U.S. MMU. For this and the next three EVAs, however, the suits were linked to Mir by the same electricity and communications/telemetry umbilical used with Orlan-DM. Valeri Polyakov (who arrived with Mir VE-3) remained sealed in the Soyuz TM-6 descent module during the EVA. The Soyuz was docked at the front of the station. Both the Mir transfer compartment and the Soyuz-TM 6 orbital module were depressurized to expand the airlock space available (the Mir base block had less airlock space than either Salyut 6 or Salyut 7). A British scientist accompanied by a British TV news crew monitored the EVA from the TsUP. Titov and Manarov left one of the transfer compartment berthing ports carrying a new detector for the TTM X-ray telescope on Kvant. The old detector was not designed for replacement, but the new one had handling aids and large fasteners easily operated using EVA gloves. The detector slid into place with difficulty, but the repair still required about an hour less than expected. Titov and Manarov then installed a special foot restraint for the Soviet-French EVA scheduled for December. The restraint was designed and manufactured on the ground using measurements they made during their February EVA.

"Mir Space Walk," *Spaceflight*, December 1988, p. 457; "Mir Mission Report: TTM Telescope Repaired," Neville Kidger, *Spaceflight*, February 1989, p. 64; "Mir Mission Report," Neville Kidger, *Spaceflight*, December 1988, p. 454; *Soviet Year in Space 1988*, Nicholas Johnson, 1989, p. 100; *Krasnaya Zvezda*, October 21, 1988, p. 3 (abstracted in *JPRS Report, Science & Technology, USSR: Space*, February 16, 1989, pp. 36-37); *Flight Crew Support on the Mir Space Station*, JSC 26898, Paul Campbell, December 1994, pp. 124-128; "U.S., Russian Suits Serve Diverse EVA Goals," James Asker, *Aviation Week & Space Technology*, January 16, 1995, pp. 40-45; "Design to Safety: Experience and Plans of the Russian Space Suit Programme," Guy Severin, *Acta Astronautica*, Vol. 32, No. 1, pp. 15-23, 1994.

November 26	Mir/Soyuz-TM 7 PE-4 launch
December 2-6	STS-27/Atlantis

December 9
1988 EVA 4
World EVA 76
French EVA 1/Russian EVA 25
Space Station EVA 33
Duration: 5:57
Spacecraft/mission: Mir PE-3/Mir PE-4
Crew: Vladimir Titov, Musa Manarov, Valeri Polyakov (Mir PE-3); Alexandr Volkov, Sergei Krikalev, Jean-Loup Chretien (Mir PE-4)
Spacewalkers: Jean-Loup Chretien, Alexandr Volkov
Purpose: Perform first French EVA; conduct engineering experiments for Hermes Development Program and Columbus space station program

This EVA made Chretien the first non-U.S./non-Soviet spacewalker. His EVA was a highlight of the 3-wk French-Soviet Aragatz mission, which began with Soyuz-TM 7's arrival at Mir on November 28. Initially the EVA was scheduled for December 12, but the TsUP elected to move it to December 9 to leave time for Chretien to participate in a second EVA if the hexagonal ERA platform did not deploy. Fully deployed, ERA measures 3.6 m (11.8 ft) wide by 3.8 m (12.5 ft) long. Chretien's EVA was scheduled to last just 3 hr. First outside, he leaned out of a transfer compartment hatch and unfolded handrails recessed into Mir's hull. Then he used springs and hooks to attach the Enchantillons space exposure rack to the handrails. It carried five technological experiments. Chretien attached electrical leads to Mir's power supply and with difficulty opened sample container lids on the rack. Volkov then joined Chretien outside to set up the 240-kg (528-lb) ERA, which included a mounting platform, deployable structure, and "filming block" for recording deployment. They attached the mounting platform to handrails on the frustum between Mir's transfer compartment and small-diameter work compartment, then attached the deployable structure. Sergei Krikalev commanded ERA to deploy from inside Mir, but it remained stubbornly folded. The frustrated cosmonauts shook the recalcitrant structure, but the TsUP rejected Volkov's offer to kick ERA. After consultation with French engineers, the TsUP told the cosmonauts to discard ERA and return inside if it failed to open by remote command. Mir then passed out of radio range, and Volkov kicked ERA several times. When the TsUP reacquired Mir, it learned that the platform was fully deployed. The cosmonauts discarded the structure and returned inside, setting a new Soviet EVA endurance record.

"Mir Mission Report," Neville Kidger, *Spaceflight*, March 1989, pp. 78-79; "France to Gain Extensive Manned Spaceflight Experience on Mir," Jeffrey Lenorovitz, *Aviation Week & Space Technology*, November 28, 1988, p. 43; *Soviet Year in Space 1988*, Nicholas Johnson, 1989, pp. 102-103; "VLD/ERA: A French Experiment on the Soviet Union Mir Station," IAA-88-050, Gilles Debas, Pierre Picard, and Patrick Aubry, 39th Congress of the International Astronautical Federation, October 15-18, 1988.

| December 21 | Mir/Soyuz-TM 6 PE-3 landing |

1989

March 13-18	STS-29/Discovery
April 27	Mir/Soyuz-TM 7 PE-4 landing
May 4-8	STS-30/Atlantis
August 8-13	STS-28/Columbia
September 5	Mir/Soyuz-TM 8 PE-5 launch
October 18-23	STS-34/Atlantis
November 22-27	STS-33/Discovery

1990

January 8
1990 EVA 1
World EVA 77
Russian EVA 26
Space Station EVA 34
Duration: 2:56
Spacecraft/mission: Mir PE-5
Crew: Alexandr Viktorenko, Alexandr Serebrov
Spacewalkers: Alexandr Viktorenko, Alexandr Serebrov
Purpose: Install new star trackers on Kvant module

The Kvant 2 module arrived at Mir's front port on December 6, 1989, and was pivoted into place at one of the radial ports using a robot arm. Viktorenko and Serebrov then moved their Soyuz-TM 8 spacecraft to the front port to make way for Progress freighters at the station's Kvant module aft port. On December 29 they began preparations for the first of five planned PE-5 EVAs to integrate Kvant 2 into the Mir complex and receive the Kristall module. On this date the cosmonauts exited Mir through one of three unoccupied radial ports in Mir's transfer compartment. The EVA's start was delayed 1 hr by an uncooperative valve - it let air escape from Soyuz-TM 8 when the cosmonauts spilled air from the transfer compartment. Viktorenko and Serebrov finally egressed during orbital night, just before midnight Moscow time. The cosmonauts had other minor problems - a broken wire in Viktorenko's suit prevented water temperature monitoring and Serebrov's coolant loop leaked. Despite these minor annoyances, the cosmonauts successfully installed two star trackers - each weighing 80 kg (176 lb) - on "standard points" on Kvant while out of contact with the TsUP. They retrieved Meduza samples from Mir's hull before ending the EVA.

The Soviet Year in Space 1990, Nicholas Johnson, 1991, pp. 98-100; TASS, January 9, 1990 (translated in *JPRS Report, Science & Technology, USSR: Space* (JPRS-USP-90-001), March 15, 1990, p. 9); TASS, December 29, 1989 (translated in *JPRS Report, Science & Technology, USSR: Space* (JPRS-USP-90-001), March 15, 1990, p. 8); "Cosmonauts Take Space Walk to Upgrade Mir Station," James Fisher, *Space News*, January 15-21, 1990, p. 4; "The Experience in Operation and Improving the Orlan-type Space Suits," I. P. Abramov, *Acta Astronautica*, Vol. 36, No. 1, July 1995, pp. 1-12.

January 9-20   STS-32/Columbia

January 11
1990 EVA 2
World EVA 78
Russian EVA 27
Space Station EVA 35
Duration: 2:54
Spacecraft/mission: Mir PE-5
Crew: Alexandr Viktorenko, Alexandr Serebrov
Spacewalkers: Alexandr Viktorenko, Alexandr Serebrov
Purpose: Modify Mir following Kvant 2 arrival; miscellaneous tasks

This EVA, preparations for which commenced immediately after the January 9 spacewalk, marked the last use of the Mir transfer compartment as an airlock until 1995, and the last use of the Orlan-DMA suit with a power umbilical. Serebrov and Viktorenko retrieved the Echantillons exposure experiment attached to Mir's hull by Jean-Loup Chretien in December 1988. They then installed space exposure cassettes containing non-metallic materials on the Mir base block and installed on Kvant the Arfa ("harp")-E experiment, which monitored the ionosphere and magneto-sphere. The cosmonauts removed the supports for the French ERA experiment from Mir's hull, then returned to the depressurized transfer compartment, where they moved the Konus #2 drogue from the +Y port (where Kvant 2 is docked) to the -Y port (opposite Kvant 2) to receive Kristall. After hard dock and activation of docking collar latches Konus #2 could be removed. Konus #1 remained in the Mir forward (-X) port at all times. Konus transfer was originally planned to occur during a separate EVA, so this reduced to four the total number of EVAs planned for Viktorenko and Serebrov.

The Soviet Year in Space 1990, Nicholas Johnson, 1991, pp. 100; TASS, January 11, 1990 (translated in *JPRS Report, Science & Technology, USSR: Space* (JPRS-USP-90-001), March 15, 1990, p. 9).

January 26
1990 EVA 3
World EVA 79
Russian EVA 28
Space Station EVA 36
Duration: 3:02
Spacecraft/mission: Mir PE-5
Crew: Alexandr Viktorenko, Alexandr Serebrov
Spacewalkers: Alexandr Viktorenko, Alexandr Serebrov
Purpose: Test Orlan-DMA without umbilical; prepare Mir for SPK tests

This EVA marked the first use of the Orlan-DMA suit with an add-on package supplying power, telemetry, and communications, rendering obsolete the umbilical used on all previous Soviet EVAs. Viktorenko and Serebrov were linked to Mir only by tethers. The Kvant 2 Special Airlock Compartment (SALC) was also inaugurated on this EVA. The SALC, which became the main airlock of the Mir complex, contained the Soviet equivalent of the U.S. MMU, the *Sredstvo Peredvizheniy Kosmonavtov* ("Cosmonaut Maneuvering Equipment") (SPK). The device was manufactured by Zvezda, the same organization which built the Orlan space suits. The cosmonauts tested the suit modifications while they installed a "dock" (recovery winch and mooring post) for the SPK outside the Kvant 2 airlock. They then removed the obstructing Kvant 2 Kurs antenna ahead of the SPK flight test. Before pulling shut Kvant 2's large EVA hatch, they installed Ferrit and Danko space exposure cassettes on the module's hull and installed the Gemma-2 camera on the Kvant 2 tracking platform.

The Soviet Year in Space 1990, Nicholas Johnson, 1991, pp. 100; TASS, January 18, 1990 (translated in *JPRS Report, Science & Technology, USSR: Space* (JPRS-USP-90-001), March 15, 1990, p. 10); TASS, January 26, 1990 (translated in *JPRS Report, Science & Technology, USSR: Space* (JPRS-USP-90-001), March 15, 1990, p. 10).

February 1
1990 EVA 4
World EVA 80
Russian EVA 29
Space Station EVA 37
Duration: 4:59
Spacecraft/mission: Mir PE-5
Crew: Alexandr Viktorenko, Alexandr Serebrov
Spacewalkers: Alexandr Viktorenko, Alexandr Serebrov
Purpose: Test the SPK cosmonaut maneuvering device

Flight Engineer Serebrov trained for many years to fly the SPK using a computer-operated simulator. For this test, he remained attached to Mir at all times by a 60-m (197-ft) tether. The slender tether was deemed necessary because the station could not maneuver to retrieve him if he became stranded out of reach by SPK failure. It was attached to an electric winch on the dock installed on the previous EVA. The winch automatically took up slack in the tether to prevent tangling. The SPK attached to "magnetic" attachment points on the dock. During the test the visitor balcony in the TsUP was packed to capacity. Serebrov flew the SPK out 5 m (16.4 ft) from Mir and back three times. He then backed to 33 m (108 ft), stopped, and completed various maneuvers. Viktorenko videotaped Serebrov. During his final test maneuvers, Serebrov determined that he was approaching the dock off course. He corrected, and the tether caused him to flip backwards and rock "like a pendulum." Despite this, Guy Severin, Zvezda chief, told the Soviet press that engineers were pleased by initial results of the SPK test.

"Get Into Your Sleighs!: A Report from the Flight Control Center," A. Tarasov, *Pravda*, February 2, 1990, p. 1 (translated in *JPRS Report, Science & Technology, USSR: Space* (JPRS-USP-90-003), July 30, 1990, pp. 14-15); "Mir Mission Report: Cosmonauts Fly Their Space Motorcycle," Neville Kidger, *Spaceflight*, July 1990, pp. 229-230; *The Soviet Year in Space 1990*, Nicholas Johnson, 1991, pp. 100-101.

February 5
1990 EVA 5
World EVA 81
Russian EVA 30
Space Station EVA 38

Duration: 3:45
Spacecraft/mission: Mir PE-5
Crew: Alexandr Viktorenko, Alexandr Serebrov
Spacewalkers: Alexandr Viktorenko, Alexandr Serebrov
Purpose: Test SPK

Before this second and final flight of the SPK, Serebrov attached the Spin-6000 device to an attachment fixture on the front of the SPK belly band. Spin-6000 measured the radiation background outside Mir, focusing on the secondary radiation produced by atomic particles striking the station's hull. Serebrov backed away to 45 m and did an "aerobatic" roll, covering a total of about 200 m. He needed help from Serebrov to redock because Spin-6000 blocked his view of the dock. According to Vladimir Shatalov, Head of Cosmonaut Training, the SPK was to be used after these first two tests for undefined "practical purposes." Other officials said that it would be used to inspect Mir's exterior. In fact, the SPK remained stored inside Mir until February 1996, when it was abandoned at the end of its dock outside the Kvant 2 airlock hatch to make more room in the Kvant 2 SALC.

"Soviet MMU Set For Tests in January," *Spaceflight*, January 1990, p. 11; "Mir Mission Report: Cosmonauts Fly Their Space Motorcycle," Neville Kidger, *Spaceflight*, July 1990, pp. 229-230; *The Soviet Year in Space 1990*, Nicholas Johnson, 1991, pp. 100-101.

February 11	**Mir/Soyuz-TM 9 PE-6 launch**
February 19	**Mir/Soyuz-TM 8 PE-5 landing**
February 28-March 4	**STS-36/Atlantis**
April 24-29	**STS-31/Discovery**

July 17
1990 EVA 6
World EVA 82
Russian EVA 31
Space Station EVA 39
Duration: 7:00
Spacecraft/mission: Mir PE-6
Crew: Anatoli Solovyov, Alexandr Balandin
Spacewalkers: Anatoli Solovyov, Alexandr Balandin
Purpose: Repair damaged thermal blankets on Soyuz TM-9

Soyuz-TM 9 arrived at Mir in tatters, with thermal blankets on its descent module flapping loosely about their forward attachment points. Without the blankets, the temperature inside the spacecraft fell, so condensation threatened to form on sensitive electronics. Temperature extremes threatened the heatshield and pyrotechnics, and the loose blankets obscured sensors used to orient the craft for reentry. After more than 150 days on Mir, PE-6 cosmonauts Solovyov and Balandin donned the Orlan-DMA space suits, entered the Kvant 2 airlock carrying ladders and tools, and depressurized the SALC to go EVA to repair Soyuz-TM 9. No EVA had been scheduled for PE-6. According to I. Vostrikov, deputy general designer at the Salyut Design Bureau, which built Kvant 2, Solovyov and Balandin then violated Kvant 2 airlock egress procedure. The cosmonauts were supposed to turn a handwheel until a 1-to-2-mm (0.04-to-0.08-in) slit opened around the lip of the hatch opening, allowing residual air to escape. Before they turned the handwheel further, releasing retaining hooks so that they could push back the hatch, they were supposed to confirm that the airlock was in vacuum using a handheld measuring device. Solovyov and Balandin turned the handwheel too far, releasing the hooks prematurely. Air pressure within the lock remained at 5 kpascal (0.74 psi) so the hatch sprang back against its hinges with a force of 400 kg (880 lb). Flight Engineer Balandin exited first. Traversing Kvant 2's 13.73-m (45-ft) length required longer than expected - about 90 min. The cosmonauts quipped after the EVA that they had plenty of handholds, but needed street signs. They rested during orbital night passes. Three hr passed before the cosmonauts finished installing a straight ladder bridging the gap between Kvant 2 and Soyuz-TM 9, and a curved ladder to the heatshield and explosive bolts. The mood in the TsUP was tense, in part because controllers could not see the cosmonauts - the TV camera cables were not long enough to reach the worksite. Solovyov and Balandin videotaped the descent module for later playback to the TsUP. The cosmonauts detected no obvious damage to the explosive bolts and heatshield. They then folded two of the thermal blankets in half, but left the third alone. By this time more than 5 hr had passed, so they hastened back to the Kvant 2 hatch, leaving ladders and tools at the worksite. They entered the Kvant 2 airlock after exceeding the 6-hr Orlan-DMA safety limit, only to find that the hatch would not close. They used the Kvant 2 instrument-science compartment (ISC) as a contingency airlock, leaving the SALC in vacuum. Soviet officials stated that the ISC could be used to extend the main airlock compartment for transferring large equipment outside the Mir station. This remains the longest Russian spacewalk to date.

Moscow Television Service, July 17, 1990. Translated in *JPRS Report, Science & Technology, USSR: Space*, October 5, 1990, (JPRS-USP-90-004), pp. 1-2; Moscow Television Service, July 18, 1990. Translated in *JPRS Report, Science & Technology, USSR: Space*, October 5, 1990 (JPRS-USP-90-004), p. 2; Moscow Domestic Service in Russian, July 19, 1990. Translated in *JPRS Report, Science & Technology, USSR: Space*, October 5, 1990 (JPRS-USP-90-004), p. 2; Translated in *JPRS Report, Science & Technology, USSR: Space*, October 5, 1990 (JPRS-USP-90-004), p. 3; "Spacewalk to Repair Damaged Hatch," Neville Kidger, *Spaceflight*, October 1990, p. 349; *The Soviet Year in Space 1990*, Nicholas Johnson, Kaman Sciences, 1991, pp. 103, 109-110. *Mir Hardware Heritage*, NASA RP-1357, David S. F. Portree, March 1995, pp. 56, 122.

July 25
1990 EVA 7
World EVA 83
Russian EVA 32
Space Station EVA 40
Duration: 3:31
Spacecraft/mission: Mir PE-6
Crew: Anatoli Solovyov, Alexandr Balandin
Spacewalkers: Anatoli Solovyov, Alexandr Balandin
Purpose: Investigate damaged Kvant 2 hatch; attempt to close and seal

Playback of the videotape made by Solovyov and Balandin and detailed post-EVA debriefing convinced engineers that Soyuz-TM 9 was in excellent condition to return to Earth. Before the cosmonauts could safely undock, however, they had to remove ladders and tools they left near Soyuz-TM 9 after their July 17 EVA. In addition, engineers wanted them to inspect the Kvant 2 hatch. A Soviet state commission authorized the cosmonauts to work outside the station for up to 9 hr if required. On this date the cosmonauts depressurized the Kvant 2 ISC and moved through the unpressurized airlock into space. First they televised images of the damaged hatch to the TsUP. One hinge was obviously deformed. Then they moved to the Soyuz-TM 9 worksite - more easily this time - and removed the ladders, stowing them on Kvant 2's hull. Meanwhile, engineers in the TsUP sought a means of closing the hatch. Despite difficulty in gaining sufficient leverage, Balandin and Solovyov forced the hatch shut. They repressurized the SALC and ISC, sealed themselves in the latter, and doffed their suits, leaving the hatch to the airlock compartment closed. After 24 hr, the external airlock hatch showed no leakage, so the TsUP gave Solovyov and Balandin permission to leave open the hatch connecting the airlock to the rest of the station. On August 4, as Balandin and Solovyov handed off Mir to the PE-7 crew of Strekalov and Manakov, Radio Moscow World Service quoted former cosmonaut Vladimir Shatalov, head of the Cosmonaut Training Center at Star City, as saying that a single EVA would be sufficient to repair the hatch. Shatalov said this in part to placate Mir's critics, who pointed out that repairs consumed much of the cosmonauts' time, decreasing time available for productive research.

Moscow Television Service, July 25, 1990. Translated in *JPRS Report, Science & Technology, USSR: Space*, October 5, 1990 (JPRS-USP-90-004), p. 3; Moscow Television Service in Russian, July 26, 1990. Translated in *JPRS Report, Science & Technology, USSR: Space*, October 5, 1990 (JPRS-USP-90-004), pp. 3-4. *Izvestiya*, July 28, 1990, p. 1. Translated in *JPRS Report, Science & Technology, USSR: Space*, October 5, 1990 (JPRS-USP-90-004), pp. 8; *The Soviet Year in Space 1990*, Nicholas Johnson, 1991, pp. 110-112; Moscow World Service in English, August 4, 1990. In *JPRS Report, Science & Technology, USSR: Space*, October 5, 1990 (JPRS-USP-90-004), p. 6.

August 1	**Mir/Soyuz-TM 10 PE-7 launch**
August 9	**Mir/Soyuz-TM 9 PE-6 landing**
October 6-10	**STS-41/Discovery**

October 29
1990 EVA 8
World EVA 84
Russian EVA 33
Space Station 41
Duration: 2:45
Spacecraft/mission: Mir PE-7
Crew: Gennadi Manakov, Gennadi Strekalov
Spacewalkers: Gennadi Manakov, Gennadi Strekalov
Purpose: Repair damaged Kvant 2 hatch

The Kvant 2 airlock hatch repair spacewalk was postponed from October 19 when Strekalov came down with a cold, but finally began late on this date. Strekalov and Manakov used a specially designed tool to remove insulation from the outside of the airlock hatch, revealing that the hinge was damaged beyond their ability to repair. They attached a special latch to ensure adequate closure and retreated inside. A scheduled EVA to prepare for transfer of solar arrays from Kristall to Kvant was postponed until after the next Principal Expedition crew could replace the hinge.

Moscow Television Service in Russian, October 30, 1990. Translated in *JPRS Report, Science & Technology, USSR: Space* (JPRS-USP-90-005), p. 6; *The Soviet Year in Space 1990*, Nicholas Johnson, Kaman Sciences, 1991, pp. 114; *Mir Hardware Heritage*, NASA RP-1357, David S. F. Portree, March 1995, pp. 127.

November 15-20	**STS-38/Atlantis**
December 2-10	**STS-35/Columbia**
December 2	**Mir/Soyuz-TM 11 PE-8 launch**
December 10	**Mir/Soyuz-TM 10 PE-7 landing**

January 7
1991 EVA 1
World EVA 85
Russian EVA 34
Space Station EVA 42
Duration: 5:18
Spacecraft/mission: Mir PE-8
Crew: Viktor Afanaseyev, Musa Manarov
Spacewalkers: Viktor Afanaseyev, Musa Manarov
Purpose: Repair damaged Kvant 2 hatch; prepare to transfer of Kristall's solar arrays to Kvant

Soyuz-TM 11 delivered Afanaseyev and Manarov, the eighth Mir Principal Expedition crew, to Mir's front port on December 4, 1990. They brought with them a replacement hinge for the Kvant 2 airlock hatch. The cosmonauts underwent a medical exam as part of their EVA preparations on January 3. On this date - Christmas Day in the Russian Orthodox religious calendar - the cosmonauts gave engineers and flight controllers in the TsUP a present by successfully replacing the hinge damaged six months earlier during PE-6. The work was described as "very complex and very delicate" because the hinge was not designed for EVA replacement. The replacement hinge was designed to be installed by weightless cosmonauts working in pressure suits with EVA tools. The EVA was scheduled to last 4 hr, 20 min. Four hr into the EVA Afanaseyev and Manarov entered the SALC and closed and sealed the hatch to check their work, then reopened it and egressed to carry out other EVA tasks. These included moving parts and equipment for the upcoming solar array transfer EVA to Kvant 2's exterior; removing a camera from the Kvant 2 "video spectrum complex" (the Gemma-2 unit for Earth environment monitoring) for repair inside Mir; and removing for return to Earth a space exposure cassette of superconductive materials.

TASS in English, December 29, 1990. In *JPRS Report, Science & Technology, USSR: Space*, February 7, 1991 (JPRS-USP-91-001), p. 12; TASS International Service in Russian, January 8, 1991. Translated in *JPRS Report, Science & Technology, USSR: Space*, February 7, 1991 (JPRS-USP-91-001), p. 13; "New Crew Launched to Mir," Neville Kidger, *Spaceflight*, March 1991, p. 96; *1991-1992 Europe & Asia in Space*, Nicholas Johnson and David Rodvold, U.S. Air Force Phillips Laboratory, 1993, pp. 62-63.

January 23
1991 EVA 2
World EVA 86
Russian EVA 35
Space Station EVA 43
Duration: 5:33
Spacecraft/mission: Mir PE-8
Crew: Viktor Afanaseyev, Musa Manarov
Spacewalkers: Viktor Afanaseyev, Musa Manarov
Purpose: Install Strela boom

Afanaseyev and Manarov installed the 45-kg (99-lb) telescoping Strela boom on a Mir base block launch shroud attachment. Strela ("arrow") was installed primarily for moving the 500-kg (1100-lb) solar arrays on Kristall to new locations on Kvant, but would also be used for moving cosmonauts and equipment around Mir's exterior and as a mobile handrail. The task was originally scheduled to occur over two EVAs. This EVA lasted almost 2 hr longer than planned, but concluded with Strela entirely installed. To test the device, Manarov rode the end of the boom while Afanaseyev operated its cranks. Before closing out the EVA the cosmonauts removed the Ferrit space exposure experiment from Kvant 2 and replaced it with the Sprut-5 device for measuring particle flow near Mir.

1991-1992 Europe & Asia in Space, Nicholas Johnson and David Rodvold, U.S. Air Force Phillips Laboratory, 1993, pp. 63; *Mir Hardware Heritage*, NASA RP 1357, David S. F. Portree, March 1995, p. 128.

January 26
1991 EVA 3
World EVA 87
Russian EVA 36
Space Station EVA 44
Duration: 6:20
Spacecraft/mission: Mir PE-8
Crew: Viktor Afanaseyev, Musa Manarov
Spacewalkers: Viktor Afanaseyev, Musa Manarov
Purpose: Install support structures on Kvant for Kristall solar arrays

The cosmonauts installed two supports for the Kristall solar arrays on either side of Kvant. They worked near the Kvant Kurs system antenna, which was used to guide Progress-M and Soyuz-TM spacecraft during docking at the Mir complex aft port. They also installed laser retroreflectors.

1991-1992 Europe & Asia in Space, Nicholas Johnson and David Rodvold, U.S. Air Force Phillips Laboratory, 1993, pp. 63; *Mir Hardware Heritage*, NASA RP 1357, David S. F. Portree, March 1995, p. 128.

April 5   **STS-37/Atlantis launch**

April 7
1991 EVA 4
World EVA 88
U.S. EVA 52
Shuttle EVA 14
Duration: 4:26
Spacecraft/mission: STS-37
Crew: Steven Nagel, Kenneth Cameron, Jerry Ross, Jerome Apt, Linda Godwin
Spacewalkers: Jerry Ross, Jerome Apt
Purpose: Deploy jammed GRO high-gain antenna

The U.S. resumed piloted spaceflight on September 29, 1988, with the launch of Discovery on mission STS-26, but no EVA was scheduled until this flight. As it turns out, the first U.S. EVA after the January 1986 Challenger accident was a contingency EVA ahead of the planned EVA. On April 6 cabin pressure aboard Atlantis was reduced to 10.2 psi ahead of a possible contingency EVA to assist in Gamma Ray Observatory (GRO) deployment and release. On this date, as GRO checkout proceeded prior to release, astronauts Jerry Ross (who participated in the last U.S. EVA in 1985) and Jay Apt checked out their EMUs, with assistance from IV crewman Kenneth Cameron and prebreathed pure oxygen for 1 hr. Linda Godwin used the RMS to lift the 15,750-kg (35,000-lb) GRO from its cradle in Atlantis' payload bay. The observatory's solar panels were commanded to open to their full span of 21 m (69 ft). The high-gain antenna unlatched, but its 5-

m (16.4-ft) boom did not deploy. Cameron and STS-37 Commander Steve Nagel helped Ross and Apt put on their suits. The crew attempted to open the high-gain by shaking GRO with Atlantis' thruster jets and the RMS. The first U.S. spacewalk since November 1985, and the first unscheduled spacewalk since April 1985, began with Ross moving down the starboard slidewire and Apt moving down the port. Seventeen min into the spacewalk, with Atlantis passing through night, Ross shoved loose the boom by exerting about 27 kg (60 lb) of force using his right hand while holding onto a GRO flight support structure trunnion with his left. The astronauts then set up a foot restraint so they could continue manual deployment, a procedure they had practiced four times in the WETF. The procedure involved removing a pin, pulling the antenna to fully deployed position, and using a wrench to lock the boom. Apt monitored Ross' movements so that he did not inadvertently damage GRO. Both astronauts had difficulty finding handholds on GRO in darkness. As Atlantis emerged into daylight, they performed some of the EVA Development Flight Experiment activities originally scheduled for April 8. They evaluated handrails (the dog bone cross section design proves superior to the round cross-section design); used the Crew Loads Instrumented Pallet (CLIP) to measure forces placed on foot restraints by simple tasks; and moved along a rope extended across the payload bay. They returned to the airlock but did not repressurize it until GRO was successfully away. Ross and Apt stuck their helmeted heads out the airlock hatch to watch GRO shrink into the distance.

STS-37 Flight Crew Report, Steven Nagel, Kenneth Cameron, Linda Godwin, Jerry Ross, and Jerome Apt, February 4, 1992, pp. 7-9; "Astronauts Give GRO a Helping Hand," Roelof Schuiling and Steven Young, *Spaceflight*, June 1991, pp. 197, 199-201; *STS-37 EVA Lessons Learned Report*, Lead EVA Flight Control Team, July 11, 1991; "Shuttle Astronauts Perform EVAs to Free Satellite, test New Hardware," James Asker, *Aviation Week & Space Technology*, April 15, 1991, p. 26; interview, David S. F. Portree with Linda Godwin, June 13, 1996; interview, David S. F. Portree with Jerry Ross, January 11, 1996..

April 8
1991 EVA 5
World EVA 89
U.S. EVA 53
Shuttle EVA 15
Duration: 5:47
Spacecraft/mission: STS-37
Crew: Steven Nagel, Kenneth Cameron, Jerry Ross, Jerome Apt, Linda Godwin
Spacewalkers: Jerry Ross, Jerome Apt
Purpose: Test CETA cart and other EVA equipment

Ross and Apt assembled a 14.6-m (46-ft) track down the port side of Atlantis' payload bay while freefloating and attached the Crew and Equipment Translation Aid (CETA) cart device. The astronauts then took turns on CETA, each placing their feet in a CETA foot restraint so that his body was parallel to the bay wall and using a handrail to pull himself the length of the track. Apt reported that "you can give yourself a couple of pulls and go all the way to the end of the bay." An electrical locomotion system with hand cranks for generating electricity for a motor and a mechanical locomotion system with a winch-type pull handle were also tested. Ross inadvertently demonstrated that the CETA parking brake did an excellent job of holding the cart in place by forgetting to turn it off before trying to move. The manual cart won out because it required less effort to operate than the electrical (second place) and mechanical versions. The 5000-series EVA gloves Ross wore proved a disappointment even though they provided superior performance on the ground. Ross, standing in the CLIP-equipped MFR on the RMS, rode far out over the main engines at varying speeds while Apt used strain gauge devices to measure when the arm's brakes slipped. Ross attempted to perform coarse positioning of the "limp RMS" to see if equipment gripped by the end effector could be positioned with power off - this did not work well. In general, RMS-based tasks took longer to perform than expected. The astronauts became cold; in their post-mission report they stated that "it is clear that the EMU has excessive cooling capacity." They warned of trouble during Space Station Freedom (SSF) assembly, when temperatures might plummet to minus 84 deg C (minus 120 deg F) if the worksite was pointed away from Earth at night. In their post-mission report, the STS-37 crew noted that:

> Over the five years since the last EVAs, much of the EVA expertise in hardware design, fabrication, and testing, EVA planning and training, flight control, and crew operations had been lost. While this mission helped to regain some of this expertise, there were also many indications that additional EVAs are required to establish the robust level of EVA capabilities that will be necessary at all levels to support assembly and operation of the SSF.

The crew recommended that at least one or two EVA missions be conducted each year to build EVA experience and aggressively flight test EVA systems ahead of SSF assembly. Despite the success of their EVAs, Ross and Apt recommended that EVAs on consecutive days be avoided.

STS-37 Flight Crew Report, Steven Nagel, Kenneth Cameron, Linda Godwin, Jerry Ross, and Jerome Apt, February 4, 1992, pp. 9-13; *STS-37 EVA Lessons Learned Report*, Lead EVA Flight Control Team, July 11, 1991; "Astronauts Give GRO a Helping Hand," Roelof Schuiling and Steven Young, *Spaceflight*, June 1991, pp. 203-204; interview, David S. FG. Portree with Linda Godwin, June 13, 1996; interview, David S. F. Portree with Jerry Ross, January 11, 1996..

| April 11 | STS-37/Atlantis landing |

April 25
1991 EVA 6
World EVA 90
Russian EVA 37
Space Station EVA 45
Duration: 2:25
Spacecraft/mission: Mir PE-8
Crew: Viktor Afanaseyev, Musa Manarov
Spacewalkers: Viktor Afanaseyev, Musa Manarov
Purpose: Inspect damaged Kurs antenna on Kvant

Progress-M 7, an automated cargo ship, approached the Mir station aft port on March 21. About 500 m (1640 ft) out, the spacecraft's Kurs automated guidance system was unable to lock on a corresponding antenna on the Kvant module, and the Progress drifted past Mir. A second docking attempt on March 23 ended with a flight controller aborting the docking after noting a "catastrophic error" in the robot ship's orientation just 20 m (65.6 ft) from the station. Progress-M 7 passed within 7 m (23 ft) of Mir, narrowly missing solar arrays and antennas. On March 26 Manarov and Afanaseyev undocked their Soyuz-TM 11 spacecraft from Mir's front port and approached the aft port using the Kurs system. Their spacecraft imitated Progress-M 7's performance, allowing them to localize the problem in the Kurs antenna on Kvant. They completed a manual docking at the aft port. Progress-M 7 docked automatically at the forward port on March 28. Afanaseyev and Manarov spent April 23 checking their Orlan-DMA suits for an EVA to inspect Kurs. On this date the EVA was delayed 15 min while Afanaseyev reconnected a cable in the Kvant 2 airlock. The cosmonauts set up an experimental thermo-mechanical joint outside Kvant 2 early in the EVA. The experiment was designed to provide data supporting Sofora truss deployment on Kvant during the next Mir Principal Expedition. Afanaseyev put the camera taken inside Mir on January 7 back into service on Kvant 2's movable platform. Meanwhile, Manarov clambered a distance of about 30 m (100 ft) to inspect the balky Kvant Kurs antenna. By doing so he violated the Soviet EVA "buddy policy" which required that Afanaseyev accompany him. Manarov televised images of the Kurs antenna to engineers in the TsUP. One 23-cm (9.2-in) parabolic dish was missing, apparently knocked off by an accidental kick during the January 26 EVA. The cosmonauts then installed markers ("road signs") on handrails to assist future EVA cosmonauts in finding their way around Mir's expanding exterior. They collected the thermomechanical joint installed at the start of EVA before returning to the Kvant 2 airlock. Manarov later received a rebuke for moving off alone to inspect the Kurs antenna.

Vremya newscast, April 25, 1991 (translated in *JPRS Report, Science & Technology, USSR: Space*, JPRS-USP-91-003, June 26, 1991, p. 1; TASS in English, in *JPRS Report, Science & Technology, USSR: Space*, JPRS-USP-91-003, June 26, 1991, p. 1; *1991-1992 Europe & Asia in Space*, Nicholas Johnson and David Rodvold, U.S. Air Force Phillips Laboratory, 1993, pp. 64-65; *Mir Hardware Heritage*, NASA RP 1357, David S. F. Portree, March 1995, p. 129; "Progress-M 7: Catastrophe Avoided," Neville Kidger, *Spaceflight*, June 1991, p. 192.

April 28-May 6	STS-39/Discovery
May 18	Mir/Soyuz-TM 12 PE-9 launch
May 26	Mir/Soyuz-TM 11 PE-8 landing
June 5-14	STS-40/Columbia

June 25
1991 EVA 7
World EVA 91
Russian EVA 38
Space Station EVA 46
Duration: 4:48
Spacecraft/mission: Mir PE-9
Crew: Anatoli Artsebarski, Sergei Krikalev
Spacewalkers: Anatoli Artsebarski, Sergei Krikalev
Purpose: Replace damaged Kurs antenna on Kvant; test new joint ahead of Sofora assembly

Progress-M 8 delivered tools and equipment for the planned 6-hr Kvant Kurs antenna repair EVA on June 1, and Artsebarski and Krikalev practiced the planned repairs inside Mir on June 14. The work was considered unusually delicate and complex because it involved small tools, such as a dental mirror, and many small parts not designed for EVA handling. In addition, there were few handholds and footholds at the Kvant work site. Getting into proper working position required a full hour when the repair was simulated in the Hydrolaboratory. Artsebarski and Krikalev rested during orbital night when visibility was too poor to permit delicate work. After repairing Kurs, the cosmonauts assembled a prototype thermo-mechanical joint outside Kvant 2 in preparation for the planned Sofora truss assembly EVAs. The joint had sleeve couplings made of titanium-nickel alloy with "memory effect," which shrank and snugged tight when heated by a hand-held heating and assembly device. Sofora was expected to be more durable than the URS truss tested on Salyut 7 (May 1986), which employed a mechanical hinged joint system. The newspaper *Izvestia* stated that: "It is sad that we have of late referred to the Mir orbital complex. . . almost exclusively in connection with repair work. No one will argue that it is [not] taking ever-increasing effort to maintain the aging orbital complex. Yet it is still not fully equipped. . ."

"Troubled Night On Board Mir," S. Leskov, *Izvestiya*, June 26, 1991, p. 1 (translated in *JPRS Report, Science & Technology, USSR: Space*, JPRS-USP-91-004, September 20, 1991, p. 2); abstract of *Vremya* TV news report, July 10, 1991, in *JPRS Report, Science & Technology, USSR: Space*, JPRS-USP-91-004, September 20, 1991, p. 3; *1991-1992 Europe & Asia in Space*, Nicholas Johnson and David Rodvold, U.S. Air Force Phillips Laboratory, 1993, pp. 67-68.

June 28
1991 EVA 8
World EVA 92
Russian EVA 39
Space Station EVA 47
Duration: 3:24
Spacecraft/mission: Mir PE-9
Crew: Anatoli Artsebarski, Sergei Krikalev
Spacewalkers: Anatoli Artsebarski, Sergei Krikalev
Purpose: Attach TREK experiment to Mir's exterior

In addition to tools for the Kvant Kurs antenna repair, Progress-M 8 delivered the 1-m (3.3-ft) TREK panel. The space exposure experiment was devised by the University of California at Berkeley to study cosmic-ray superheavy nuclei by recording their tracks through layers of phosphate glass. TREK was designed to remain outside on Kvant 2 for 2 yr, then be recovered and returned to Earth for analysis. Artsebarski and Krikalev also installed charged particle detectors, retrieved the thermomechanical joint assembled during the previous EVA, and tested a new TV camera. They used the Strela boom to move around Mir's exterior and completed the EVA 2 hr ahead of schedule.

"Mir Mission Report: Cosmonauts Chalk Up More than Thirty Hours of Spacewalks," Neville Kidger, *Spaceflight*, October 1991, pp. 359-360; transcript of *Man, Earth, Universe* program, Moscow Central TV Second Program, July 20, 1991 (translated in *JPRS Report, Science & Technology, USSR: Space*, JPRS-USP-91-004, September 20, 1991, pp. 4-5).

July 15
1991 EVA 9
World EVA 93
Russian EVA 40
Space Station EVA 48
Duration: 5:45
Spacecraft/mission: Mir PE-9
Crew: Anatoli Artsebarski, Sergei Krikalev
Spacewalkers: Anatoli Artsebarski, Sergei Krikalev
Purpose: Prepare worksite for Sofora girder construction

This was the first of four planned EVAs dedicated to Sofora truss assembly. Sofora, named for a fast-growing central Asian shrub, was developed by NPO Energia. During the EVA, Artsebarski noted unusually heavy air leakage through abrasions in his gloves. This was the eleventh EVA for his suit. The cosmonauts used Strela to move themselves and the Sofora mounting platform from Kvant 2 to the worksite on Kvant. They then attached four heating and assembly devices to exterior electrical power outlets.

TASS in English, July 19, 1991 (transcribed in *JPRS Report, Science & Technology, USSR: Space*, JPRS-USP-91-004, September 20, 1991, p. 4); transcript of Moscow All-Union First Program Network, July 27, 1991 (translated in *JPRS Report, Science & Technology, USSR: Space*, JPRS-USP-91-004, September 20,

1991, p. 6); "Goodbye, Spacesuit? Report from the Flight Control Center," A. Tarasov, *Pravda*, July 29, 1991, p. 2 (translated in *JPRS Report, Science & Technology, USSR: Space*, JPRS-USP-91-007, November 20, 1991, pp. 5-7); *1991-1992 Europe & Asia in Space*, Nicholas Johnson and David Rodvold, U.S. Air Force Phillips Laboratory, 1993, pp. 67-68; "The Experience in Operation and Improving the Orlan-type Space Suits," I. P. Abramov, *Acta Astronautica*, Vol. 36, No. 1, July 1995, pp. 1-12.

July 19

1991 EVA 10
World EVA 94
Russian EVA 41
Space Station EVA 49
Duration: 5:28
Spacecraft/mission: Mir PE-9
Crew: Anatoli Artsebarski, Sergei Krikalev
Spacewalkers: Anatoli Artsebarski, Sergei Krikalev
Purpose: Start assembling Sofora girder

Flight Engineer Sergei Krikalev left the Kvant 2 SALC, moved down the module to the base block, and took up the controls of the Strela boom, which he used to transfer Artsebarski and two boxes of Sofora parts to the Kvant worksite. He also transferred the first cubical, half-meter-wide Sofora truss section, which the cosmonauts assembled inside Mir before the EVA to serve as a base for the remaining 20 truss sections. The cosmonauts attached the mounting platform moved on the previous EVA to Kvant's hull, then began Sofora assembly. The truss was put together lying back over Soyuz-TM 12 at the aft port, parallel to the long axis of the Mir base block. Krikalev and Artsebarski used the four heating and assembly devices to shrink the memory metal sleeves in the truss joints. They had difficulty seeing their work as the lighting changed, but managed to keep working during orbital night. The cosmonauts were unable to use foot restraints provided because the distance between the restraints and their work was different than on Earth, so they relied on their hands and arms to hold position. They recorded their operations on video-tape, then transmitted the recordings to specialists in the TsUP during communication sessions after the EVA. Krikalev and Artsebarski assembled three Sofora segments before closing out this, their fourth EVA together.

TASS in English, July 19, 1991, in *JPRS Report, Science & Technology, USSR: Space*, JPRS-USP-91-004, September 20, 1991, p. 4; transcript of *Man, Earth, Universe* program, Moscow Central TV Second Program, July 20, 1991 (translated in *JPRS Report, Science & Technology, USSR: Space*, JPRS-USP-91-004, September 20, 1991, pp. 4-5); "Goodbye, Spacesuit? Report from the Flight Control Center," A. Tarasov, *Pravda*, July 29, 1991, p. 2 (translated in *JPRS Report, Science & Technology, USSR: Space*, JPRS-USP-91-007, November 20, 1991, pp. 5-7); *1991-1992 Europe & Asia in Space*, Nicholas Johnson and David Rodvold, U.S. Air Force Phillips Laboratory, 1993, p. 68.

July 23

1991 EVA 11
World EVA 95
Russian EVA 42
Space Station EVA 50
Duration: 5:34
Spacecraft/mission: Mir PE-9
Crew: Anatoli Artsebarski, Sergei Krikalev
Spacewalkers: Anatoli Artsebarski, Sergei Krikalev
Purpose: Continue Sofora assembly

The cosmonauts partly assembled Sofora segments inside Mir between EVAs to save time. Artsebarski's liquid cooling garment connector came apart during suit checkout, probably because its operational lifetime had been exceeded. During this EVA they added 11 more segments to Sofora, commenting on how easy it was to assemble. After the EVA, veteran cosmonaut V. I. Sevastyanov, host of a popular Moscow TV science program, told his viewers that the EVA was performed quickly and smoothly because of intense preparation and training. He added that, "a spacewalk is like a stage performance. And how much work is necessary backstage, before [the curtain goes up]?"

Transcript of *Man, Earth, Universe* program, Moscow Central TV Second Program, July 20, 1991 (translated in *JPRS Report, Science & Technology, USSR: Space*, JPRS-USP-91-004, September 20, 1991, pp. 4-5); *1991-1992 Europe & Asia in Space*, Nicholas Johnson and David Rodvold, U.S. Air Force Phillips Laboratory, 1993, p. 68; "The Experience in Operation and Improving the Orlan-type Space Suits," I. P. Abramov, *Acta Astronautica*, Vol. 36, No. 1, July 1995, pp. 1-12.

July 27
1991 EVA 12
World EVA 96
Russian EVA 43
Space Station EVA 51
Duration: 6:49
Spacecraft/mission: Mir PE-9
Crew: Anatoli Artsebarski, Sergei Krikalev
Spacewalkers: Anatoli Artsebarski, Sergei Krikalev
Purpose: Complete Sofora assembly

PE-9's sixth and final EVA began with release into space of the worn-out Orlan-DMA #10 suit. The newspaper *Pravda* lamented the suit's disposal in space, saying that it might have been returned to Earth and sold for profit to a museum. The suit was worn 9 times by different cosmonauts. The three remaining Sofora segments were assembled, then the truss was attached to its mounting platform on Kvant and raised so that it was nearly perpendicular to the long axis of the Mir base block. Sofora was sloped 11 deg toward Mir's front to place its top above the station's center-of-gravity. Artsebarski climbed to the top of the truss and attached a Soviet flag mounted in a metal frame. Moscow TV's *Vremya* news program stated that, "it is not difficult to understand Anatoli Artsebarski and Sergei Krikalev, who, on their own initiative, placed a Soviet flag atop the girder. After all, our country has not totally fallen apart yet and there are still things which we do better than anyone else in the world." Artsebarski's helmet visor fogged up because his suit's heat exchanger ran out of water, so Krikalev had to guide him back to the Kvant 2 SALC. According to cosmonaut Sevastyanov, speaking as host of a TV science program, Sofora was as tall as a five-story building and would "be subjected to burning frosts and radiation and. . . left alone for a whole year with only its sensors for company." If it proved able to withstand these conditions, Sevastyanov reported, a thruster package for Mir roll control would be placed on top. Sofora had an attachment point on top for receiving the thruster package, and was hinged a third of the way up so its top could be bent down and placed within easy reach. The cosmonauts reported bruises on their hands, elbows, and shoulders after the EVA.

Transcript of *Man, Earth, Universe* program, Moscow Central TV Second Program, July 20, 1991 (translated in *JPRS Report, Science & Technology, USSR: Space*, JPRS-USP-91-004, September 20, 1991, pp. 4-5); transcript of *Vremya* newscast , July 23, 1991 (translated in *JPRS Report, Science & Technology, USSR: Space*, JPRS-USP-91-004, September 20, 1991, p. 6); transcript of *Vremya* newscast, July 27, 1991 (translated in *JPRS Report, Science & Technology, USSR: Space*, JPRS-USP-91-004, September 20, 1991, p. 7); "Goodbye, Spacesuit? Report from the Flight Control Center," A. Tarasov, *Pravda*, July 29, 1991, p. 2 (translated in *JPRS Report, Science & Technology, USSR: Space*, JPRS-USP-91-007, November 20,

1991, pp. 5-7); *1991-1992 Europe & Asia in Space*, Nicholas Johnson and David Rodvold, U.S. Air Force Phillips Laboratory, 1993, p. 68; "Cosmonaut Rescued from Atop Mir Tower During Station EVA," *Aviation Week & Space Technology*, August 5, 1991, p. 20; "The Experience in Operation and Improving the Orlan-type Space Suits," I. P. Abramov, *Acta Astronautica*, Vol. 36, No. 1, July 1995, pp. 1-12.

August 2-11	STS-43/Atlantis
September 12-18	STS-48/Discovery
October 2	Mir/Soyuz-TM 13 PE-10/VE-4 launch
October 10	Mir/Soyuz-TM 12 PE-9/VE-4 landing
November 24-December 1	STS-44/Atlantis

1992

January 22-30	STS-42/Discovery

February 20
1992 EVA 1
World EVA 97
Russian EVA 44
Space Station EVA 52
Duration: 4:12
Spacecraft/mission: Mir PE-10
Crew: Alexandr Volkov, Sergei Krikalev
Spacewalkers: Alexandr Volkov, Sergei Krikalev
Purpose: Perform miscellaneous maintenance tasks

Flight Engineer Sergei Krikalev was to have returned to Earth with Anatoli Artsebarski in October, but the Soviet Union's collapse following August's failed *coup d'etat* against Mikhail Gorbachev meant a shortage of launch vehicles and a need to woo independent Kazakstan, where Russia's Baikonur launch facility is located. Mission planners thus combined two missions, replacing the flight engineer from Soyuz-TM 13 with Kazakh cosmonaut-researcher Toktar Aubakirov. The Soyuz-TM 13 crew consisted also of veteran cosmonaut Alexandr Volkov and an Austrian cosmonaut-researcher. With no new flight engineer to replace him, Krikalev had to stay aboard Mir for an additional 6 mo, giving him the opportunity to add to his career total of six EVAs. Volkov's suit heat exchanger clogged at the start of this EVA, forcing him to remain near the Kvant 2 SALC so he could use the module's heat exchanger. Russian sources stated later that the malfunction occurred because the suit was stored for several months. Despite being tied by his cooling umbilical to the airlock, Volkov assisted with installation of space exposure experiments on Kvant 2. Then Krikalev moved off to carry out the remaining EVA tasks by himself, an obvious violation of the Russian EVA "buddy system." According to Nikolai Yuzov, head of the Star City space training department, EVA cost 100,000 roubles/hr, which encouraged cosmonauts and ground controllers to attempt to complete EVAs once they were started despite risks. Krikalev removed Sofora assembly equipment and cleaned cameras on Kvant, then collected an experimental solar array section that was added to the base block top array in 1988. The EVA required less time than planned despite Volkov's absence. With the conclusion of this EVA, Krikalev established a new world record for total EVA time of 36 hr, 29 min, which stood for more than 4 yr. Volkov's recalcitrant suit was later released into space, a move Krikalev condemned, saying that it might have been returned to Earth and sold at auction.

1991-1992 Europe & Asia in Space, Nicholas Johnson and David Rodvold, U.S. Air Force Phillips Laboratory, 1993, p. 71; "Mir Cosmonauts Continue Work," Neville Kidger, *Spaceflight*, April 1992, pp. 120; "The Experience in Operation and Improving the Orlan-type Space Suits," I. P. Abramov, *Acta Astronautica*, Vol. 36, No. 1, July 1995, pp. 1-12.

March 17	**Mir/Soyuz-TM-14 PE-11 launch**
March 24-April 2	**STS-45/Atlantis**
March 25	**Mir/Soyuz-TM 13 PE-10 landing**
May 7	**STS-49/Endeavour launch**

May 10
1992 EVA 2
World EVA 98
U.S. EVA 54
Shuttle EVA 16
Duration: 3:43
Spacecraft/mission: STS-49
Crew: Daniel Brandenstein, Kevin Chilton, Pierre Thuot, Kathy Thornton, Richard Hieb, Thomas Akers, Bruce Melnick
Spacewalkers: Pierre Thuot, Richard Hieb
Purpose: Retrieve Intelsat VI satellite and install perigee kick motor

The Shuttle orbiter Endeavour began its career in the spotlight with three planned EVAs, a record for a single Shuttle flight. On May 8 the crew, commanded by Daniel Brandenstein, lowered Endeavour's cabin pressure to 70.3 kpascal (10.2 psi) to reduce prebreathe time. They also checked the four EMUs carried on board and tested the RMS. On this date Rick Hieb and Pierre Thuot suited up with assistance from Thomas Akers, who, with Kathy Thornton, was scheduled to perform the second EVA of the flight, during which the Assembly of Station by EVA Methods (ASEM) experiment would be performed. Hieb and Thuot were charged with capturing the Intelsat VI satellite, stranded in low Earth orbit since March 14, 1990, and attaching a 10,455-kg (23,000-lb) solid-fueled perigee kick motor to boost it to its proper place in geosynchronous orbit. They entered Endeavour's airlock while Intelsat VI was still 13 km (8 mi) away. Thuot, who held a capture bar device equipped with an RMS grapple fixture, rode the end of the RMS, which was operated by Bruce Melnick. During orbital night Endeavour completed rendezvous and Thuot attempted to attach the capture bar to the 5.2-m-by-3.2-m (17-ft-by-12-ft) satellite. Differences between ground training and actual orbital tasks thwarted his efforts and Intelsat VI began to oscillate. Several times the RMS stopped moving because it was driven into positions its joints could not support. After three capture attempts Endeavour moved off to permit Intelsat controllers to damp its wobble.

"Endeavour's Arduous Maiden Voyage," Roelof Schuiling, *Spaceflight*, August 1993, pp. 272; "Intelsat Rescue, Space Station EVAs Set for First Endeavour Flight Test," Craig Covault, *Aviation Week & Space Technology*, May 4, 1992, p. 70-74; "Endeavour's Intelsat Rescue Sets EVA, Rendezvous Records," Craig Covault, *Aviation Week & Space Technology*, May 18, 1992, pp. 22-26; interview, David S. F. Portree with Kathy Thornton, June 17, 1996.

May 11
1992 EVA 3
World EVA 99

U.S. EVA 55
Shuttle EVA 17
Duration: 5:30
Crew: Daniel Brandenstein, Kevin Chilton, Pierre Thuot, Kathy Thornton, Richard Hieb, Thomas Akers, Bruce Melnick
Spacewalkers: Pierre Thuot, Richard Hieb
Purpose: Retrieve Intelsat VI satellite and install perigee kick motor

NASA again tried the original plan for capturing Intelsat VI, this time with greater care taken in positioning Thuot and less force on the capture bar. Endeavour again completed rendezvous in darkness, but Thuot waited until orbital sunrise to make another attempt. He tried five more times to attach the capture bar while Melnick operated the RMS and Hieb stood by in the payload bay. Although his alignment was unquestionably correct, the bar refused to seat and the satellite began wobbling again. Endeavour backed away to allow Intelsat controllers to stabilize it. Thuot later remarked that handling Intelsat VI "was much more dynamic than our training had led us to believe."

"Endeavour's Arduous Maiden Voyage," Roelof Schuiling, *Spaceflight*, August 1993, p. 272; "Pierre Thuot Speaks About Astronauts 'On the Job'," Ben Evans, *Spaceflight*, February 1994, p. 48; "Endeavour's Intelsat Rescue Sets EVA, Rendezvous Records," Craig Covault, *Aviation Week & Space Technology*, May 18, 1992, pp. 22-26.

May 13
1992 EVA 4
World EVA 100
U.S. EVA 56
Shuttle EVA 18
Duration: 8:29
Crew: Daniel Brandenstein, Kevin Chilton, Pierre Thuot, Kathy Thornton, Richard Hieb, Thomas Akers, Bruce Melnick
Spacewalkers: Pierre Thuot, Richard Hieb, Thomas Akers
Purpose: Retrieve Intelsat VI satellite and install perigee kick motor

While controllers on the ground determined if Endeavour had enough propellant to carry out a third rendezvous, the crew proposed capturing the satellite using a three-person EVA (the first in history) and components of the ASEM experiment payload. The ASEM struts would be assembled into a triangular structure to which the three EVA astronauts could attach their feet. Endeavour would maneuver under Intelsat VI and the astronauts would grasp the satellite with their hands. Astronauts Story Musgrave, Richard Clifford, and James Voss used the WETF to test whether three astronauts could fit in the airlock at once and to determine positioning in the payload bay for the capture. Meanwhile, the Intelsat organization verified that the satellite's surface temperature would not exceed the 160 deg C (320 deg F) glove touch temperature limit, and engineers from Hughes, the maker of the satellite, determined the best grab locations. At one point on May 12, seven EMUs were being used simultaneously: three in Endeavour's airlock, three in the WETF, and one in a NASA JSC vacuum chamber. On this date, Kathy Thornton helped Akers, Hieb, and Thuot suit up, and the astronauts commenced the 100th EVA in history. By the time they were finished, it was also the longest EVA in history. With Thuot on the RMS, Hieb near the starboard payload bay wall, and Akers in the center of the bay attached to an ASEM strut, Brandenstein edged Endeavour toward the satellite's underside. The astronauts studied the 14,400-kg (32,000-lb) satellite's slow rotation for about 15 min, then together grasped it. Hieb attached the capture bar while Thuot and Akers held Intelsat VI, then Melnick grappled the bar with the RMS to move the satellite into position above the perigee kick motor. The astronauts attached the motor, then retreated to the airlock while Kathy Thornton activated springs to propel Intelsat VI out of the payload bay. After two failed attempts Intelsat VI was sent on its way. The three astronauts returned to the payload bay and cleaned up, stowing foot restraints and a camera. On May 15, the perigee kick motor fired to begin Intelsat VI's long-delayed voyage to its assigned slot in geosynchronous orbit.

"Endeavour's Arduous Maiden Voyage," Roelof Schuiling, *Spaceflight*, August 1993, pp. 272-273; "Pierre Thuot Speaks About Astronauts 'On the Job'," Ben Evans, *Spaceflight*, February 1994, p. 48; "Astronauts, Engineers, and Simulators Mobilized for Satellite Rescue," *Aviation Week & Space Technology*, May 18, 1992, p. 25; "Endeavour's Intelsat Rescue Sets EVA, Rendezvous Records," Craig Covault, *Aviation Week & Space Technology*, May 18, 1992, pp. 22-26; interview, David S. F. Portree with Kathy Thornton, June 17, 1996.

May 14
1992 EVA 5
World EVA 101
U.S. EVA 57
Shuttle EVA 19
Duration: 7:44
Crew: Daniel Brandenstein, Kevin Chilton, Pierre Thuot, Kathy Thornton, Richard Hieb, Thomas Akers, Bruce Melnick
Spacewalkers: Kathy Thornton, Thomas Akers
Purpose: Test equipment for Space Station Freedom Program

The fourth EVA of Endeavour's first flight was dedicated to the ASEM experiment. ASEM was to have been the focus of the mission's second and third EVAs. The ASEM experiment was built by McDonnell-Douglas, Space Station Freedom truss prime contractor. When manifested, ASEM was an exercise in assembling the Freedom truss. At about the time ASEM was manifested, however, NASA took the decision to launch the truss in pre-assembled, pre-integrated segments. ASEM, according to Thornton, became "an exercise in frustration" involving excessive "arm work" and freefloating. Thornton and Akers assembled a pyramidal structure 4.6 m (15 ft) wide from struts and connectors ("sticks and balls") carried in the Mission Peculiar Equipment Support Structure (MPESS) in the payload bay. Because of SSF engineering requirements, ASEM struts and connectors were not designed for optimum EVA handling. For example, the narrow "necks" on the ends of the beams which made handling relatively easy during the STS 61-B EASE/ ACCESS experiments were not included because they weakened the truss. Thuot monitored Thornton and Akers and kept track of the abundant hardware in the payload bay from Endeavour's flight deck. After they completed the ASEM structure, Melnick used the RMS to position it so the spacewalkers could attach the MPESS. According to *Aviation Week & Space Technology* magazine, this exercise was "directly applicable to planning for attachment of space station modular 'nodes' to the preintegrated station truss on Shuttle missions starting in 1995-96." Akers and Thornton fell behind the EVA timeline, so a plan to have each astronaut ride the MPESS as Melnick hoisted it over Endeavour's nose was abandoned. The exercise was important because the over-the-nose location was an important station assembly area. Mass-handling exercises using the 1725-kg (3800-lb) MPESS with application to station Orbital Replaceable Unit (ORU) installation were also truncated because of time pressure and because the Intelsat VI retrieval provided needed data. The EVA included the Crew Self Rescue (CSR) flight demonstration, which tested equipment to allow an EVA astronaut to safely return to a space station if he or she became untethered. Six self-rescue devices were included, but the extra time spent on ASEM meant that only the Crew Propulsive Device (CPD) could be tested. The CPD was a hand-held nozzle assembly resembling the Gemini HHMU which was fed by a compressed nitrogen tank mounted on the PLSS. The CPD worked as expected, but time constraints meant that it could be evaluated for only about 10 min of a planned 15 to 20 min. The other CSR devices included inflatable and telescoping poles and a rope "bola" device the drifting astronaut could throw to hook to a station strut. The devices were all tested before flight by suited astronauts aboard the KC-135 aircraft at NASA JSC. During the EVA, Thornton's EMU provided improper display data and Endeavour's Ku-band antenna stopped operating, limiting TV coverage.

STS-49 Crew Self Rescue Media Guide, NASA JSC, April 1992; "Endeavour's Arduous Maiden Voyage," Roelof Schuiling, *Spaceflight*, August 1993, p. 273; "Pierre Thuot Speaks About Astronauts 'On the Job'," Ben Evans, *Spaceflight*, February 1994, p. 48; "EVAs to Influence Development of Space Station Hardware," *Aviation Week & Space Technology*, May 18, 1996, pp. 25-26. interview, David S. F. Portree with Kathy Thornton, June 17, 1996; interview, David S. F. Portree with Steve Glenn, August 21, 1996.

May 16	**STS-49/Endeavour landing**
June 25-July 9	**STS-50/Columbia**

July 8
1992 EVA 6
World EVA 102
Russian EVA 45
Space Station EVA 53
Duration: 2:03
Spacecraft/mission: Mir PE-11
Crew: Alexandr Viktorenko, Alexandr Kaleri
Spacewalkers: Alexandr Viktorenko, Alexandr Kaleri
Purpose: Inspect Kvant 2 gyrodynes; evaluate difficulty of gyrodyne repair

This EVA, the only one scheduled for Mir PE-11, was planned to last 1 hr, 55 min. Viktorenko and Kaleri used large shears to cut through thermal insulation on Kvant 2 module to reach its gyrodynes, electrically driven gyroscopes which stabilize and maneuver the Mir complex without using propellant. Four of the six gyrodynes launched on Kvant 2 had ceased to operate. One inside Kvant also failed, but this was replaceable without an EVA. The cosmonauts inspected and televised the gyrodynes for engineers in Kaliningrad. Before closing out the EVA, they tested binoculars compatible with a space suit visor for inspecting Mir's outlying areas.

"French Cosmonaut Visits Mir," Neville Kidger, *Spaceflight*, November 1992, p. 361; *1991-1992 Europe & Asia in Space*, Nicholas Johnson and David Rodvold, U.S. Air Force Phillips Laboratory, 1993, p. 73.

July 27	**Mir/Soyuz-TM 15 PE-12 launch**
July 31-August 8	**STS-46/Atlantis**
August 10	**Mir/Soyuz-TM 14 PE-11 landing**

September 3
1992 EVA 7
World EVA 103
Russian EVA 46
Space Station EVA 54
Duration: 3:56
Spacecraft/mission: Mir PE-12
Crew: Sergei Avdeyev, Anatoli Solovyov

Spacewalkers: Sergei Avdeyev, Anatoli Solovyov
Purpose: Prepare worksite for VDU installation atop Sofora truss

Almost as soon as PE-12 cosmonauts Avdeyev and Solovyov took charge of Mir from Viktorenko and Kaleri they began preparations for a series of EVAs to install the 700-kg (1549-lb) VDU thruster package on top of the Sofora truss on the Kvant module. On August 18 the VDU arrived at Mir's aft port aboard Progress-M 14 in a special compartment replacing the normal Progress-M fluid cargo compartment. On September 2 the TsUP commanded Progress-M 14 to "unload" the VDU. On this EVA Avdeyev and Solovyov installed a locking device on Sofora to hold the truss securely while bent back. They used the Strela boom to move themselves and their equipment about the station.

"French Cosmonaut Visits Mir," Neville Kidger, *Spaceflight*, November 1992, p. 362; *1991-1992 Europe & Asia in Space*, Nicholas Johnson and David Rodvold, U.S. Air Force Phillips Laboratory, 1993, p. 74; "Russia to Upgrade Mir 1 Space Station, Prepares for New Orbital Facility," Jeffrey Lenorovitz, *Aviation Week & Space Technology*, May 4, 1992, p. 84.

September 7
1992 EVA 8
World EVA 104
Russian EVA 47
Space Station EVA 55
Duration: 5:08
Spacecraft/mission: Mir PE-12
Crew: Sergei Avdeyev, Anatoli Solovyov
Spacewalkers: Sergei Avdeyev, Anatoli Solovyov
Purpose: Install VDU atop Sofora truss

Avdeyev and Solovyov bent back Sofora on the hinge a third of the way along its length and locked it into position to receive the VDU. To ease installation, the thruster package deployed from Progress-M 14 at an angle matching the top of the bent-back Sofora truss. They laid a 14-m (46-ft) power cable along the truss and attached metal braces to the VDU for securing it to Sofora. Working by flashlight during orbital night, they removed the metal frame containing the tattered remnants of the Soviet flag placed atop Sofora in 1991. Ground stations of independent Ukraine suspended service during the EVA, severely limiting communications between TsUP and cosmonauts.

"French Cosmonaut Visits Mir," Neville Kidger, *Spaceflight*, November 1992, p. 362; *1991-1992 Europe & Asia in Space*, Nicholas Johnson and David Rodvold, U.S. Air Force Phillips Laboratory, 1993, p. 74.

September 11
1992 EVA 9
World EVA 105
Russian EVA 48
Space Station EVA 56
Duration: 5:44
Spacecraft/mission: Mir PE-12
Crew: Sergei Avdeyev, Anatoli Solovyov
Spacewalkers: Sergei Avdeyev, Anatoli Solovyov
Purpose: Install VDU atop Sofora truss

The cosmonauts attached the VDU to Sofora's top and straightened the truss, then completed electrical connections between the VDU and Mir. VDU installation was originally scheduled to require four spacewalks, but Avdeyev and Solovyov finished the work in three.

"French Cosmonaut Visits Mir," Neville Kidger, *Spaceflight*, November 1992, p. 363; *1991-1992 Europe & Asia in Space*, Nicholas Johnson and David Rodvold, U.S. Air Force Phillips Laboratory, 1993, p. 74.

September 15
1992 EVA 10
World EVA 106
Russian EVA 49
Space Station EVA 57
Duration: 3:33
Spacecraft/mission: Mir PE-12
Crew: Sergei Avdeyev, Anatoli Solovyov
Spacewalkers: Sergei Avdeyev, Anatoli Solovyov
Purpose: Retrieve space exposure samples; move Kurs unit on Kristall

In this final EVA of Mir PE-12, Avdeyev and Solovyov moved the Kurs rendezvous and docking system antenna on Kristall to permit Soyuz-TM 16 to dock at the Kristall androgynous docking unit, certifying the unit ahead of a planned docking by the U.S. Space Shuttle. Before returning inside Avdeyev and Solovyov removed from the base block top array an experimental solar panel that had been exposed to space for 4 years. They also removed micrometeorite panels and samples of construction materials from Kvant 2's exterior for return to Earth.

"French Cosmonaut Visits Mir," Neville Kidger, *Spaceflight*, November 1992, p. 363; *1991-1992 Europe & Asia in Space*, Nicholas Johnson and David Rodvold, U.S. Air Force Phillips Laboratory, 1993, p. 74.

September 12-20	STS-47/Endeavour
October 22-November 1	STS-52/Columbia
December 2-9	STS-53/Discovery

1993

January 13	STS-54/Endeavour launch

January 17
1993 EVA 1
World EVA 107
U.S. EVA 58
Shuttle EVA 20
Duration: 4:28
Spacecraft/mission: STS-54
Crew: John Casper, Donald McMonagle, Mario Runco, Gregory Harbaugh, Susan Helms
Spacewalkers: Gregory Harbaugh, Mario Runco
Purpose: Gain EVA experience ahead of SSF assembly

In a 1993 interview, STS-49 astronaut Pierre Thuot stated that," I believe we need to increase our experience level amongst the Astronaut Office as well amongst the Johnson Space Center EVA team, since we've only had two missions with spacewalks since we resumed flying in 1988; I think it is very important that we should fly more missions to prepare us for Space Station." Thuot and his shipmates experienced difficulties during the STS-49 EVAs which cast a shadow over planned SSF assembly. The NASA EVA community took note of this, and proposed a series of Development Test Objective (DTO) 1210 (EVA Operations Procedures/Training) EVAs. The broad objectives of the DTO 1210 EVAs were to

* broaden EVA knowledge through planning and practice

* apply knowledge gained to future EVAs

* better quantify human performance

* refine the EVA timelining process

* evaluate Hubble Space Telescope (HST) and Space Station hardware

For this first DTO 1210 EVA, Endeavour's crew depressurized the cabin to 10.2 psi on January 15 to reduce prebreathe time. Harbaugh and Runco checked out their suits on January 16. Harbaugh was the HST EVA backup astronaut, the first backup astronaut of any kind in the history of the Shuttle program. On this date Harbaugh and Runco conducted a 40-min prebreathe, then left the airlock 40 min late because preparations took longer than expected. Susan Helms served as IV crewmember supporting the EVA. No extra equipment was added to Endeavour's manifest for this EVA. Harbaugh and Runco tested carrying a large object by carrying each other; demonstrated large tool use with a tool for manually positioning the Tracking and Data Relay Satellite tilt table; and tested their ability to align bulky objects by placing each other in the bracket which holds the EMU in the airlock. The EVA was videotaped for study by EVA operations engineers on the ground. The astronauts had to close out the EVA at the planned time despite their late start because it was assigned a lower priority than observations by one of Endeavour's payloads, the Diffuse X-ray Spectrometer, which had to be suspended during the EVA. After returning to the cabin Runco and Harbaugh recorded answers to detailed EVA questions. After they returned to Houston, they repeated their STS-54 tasks in the WETF to help improve EVA training.

"Pierre Thuot Speaks About Astronauts 'On the Job'," *Spaceflight*, February 1994, p. 48; *STS-54 Space Shuttle Mission Report*, NSTS-08282, March 1993, pp. 3; "Endeavour Advances Shuttle Capabilities, Astrophysics," James Asker, *Aviation Week & Space Technology*, January 25, 1993, p. 37; "EVA DTO 1210 Results," presentation materials, Thomas Doeling, August 12, 1994.

January 19	**STS-54/Endeavour landing**
January 24	**Mir/Soyuz-TM 16 PE-13 launch**
February 1	**Mir/Soyuz-TM 15 PE-12 landing**
April 8-17	**STS-56/Discovery**

April 19
1993 EVA 2

World EVA 108
Russian EVA 50
Space Station EVA 58
Duration: 5:25
Spacecraft/mission: Mir PE-13
Crew: Gennadi Manakov, Alexandr Poleshchuk
Spacewalkers: Gennadi Manakov, Alexandr Poleshchuk
Purpose: Use Strela boom to move cosmonauts and solar array drive motor to worksite on Kvant; install first array drive for Kristall arrays on Kvant.

Poleshchuk and Manakov stepped outside ahead of schedule as Mir passed over China. For the first time the cosmonauts worked on a contractual basis; one source reported that they were paid one million roubles for three EVAs. Flight Engineer Poleshchuk moved to the Strela controls on the base block while Manakov took up position on the boom's end, and Poleshchuk moved him to the worksite on Kvant. They then used Strela to transfer a solar array electric drive. By the start of the third hr of the EVA, the cosmonauts moved one container and slipped 10 min in their schedule. Telemetry indicated that Poleschuk's suit ventilation was not operating properly. With some difficulty the cosmonauts attached one drive to the framework installed on Kvant in 1991, then connected plugs to link it to Mir's electrical supply. Poleshchuk then discovered that one of Strela's two control handles had come off and floated away from Mir. The planned April 23 EVA had to be postponed until after about May 20, when the next freighter, Progress-M 18, would deliver a new handle. Solovyov said after the EVA that, "we will be sure to screw the handle on tighter next time."

"High Hopes for Mir's Earning Power Dashed," Peter de Selding, *Space News*, May 10-16, 1993, pp. 3, 21; "Mir Enters Eighth Year in Orbit," Neville Kidger, *Spaceflight*, September 1993, pp. 317-318.

April 25-May 6   **STS-55/Columbia**

June 18
1993 EVA 3
World EVA 109
Russian EVA 51
Space Station EVA 59
Duration: 4:33
Spacecraft/mission: Mir PE-13
Crew: Gennadi Manakov, Alexandr Poleshchuk
Spacewalkers: Gennadi Manakov, Alexandr Poleshchuk
Purpose: Repair Strela boom; install second solar array drive on Kvant

Progress-M 18 delivered a replacement Strela handle on May 24. PE-13's second EVA was scheduled to last 5 hr. The cosmonauts installed the new Strela handle, then used the boom to move the second solar drive container to the worksite on Kvant. In contrast to their April 19 EVA, Manakov and Poleshchuk installed the second drive with few problems. In fact, the cosmonauts completed the installation ahead of schedule, so they were able to spend several minutes televising images of Mir's exterior to engineers in the TsUP.

"Mir Enters Eighth Year in Orbit," Neville Kidger, *Spaceflight*, September 1993, p. 318.

June 21   **STS-57/Endeavour launch**

June 25
1993 EVA 4
World EVA 110
U.S. EVA 59
Shuttle EVA 21
Duration: 5:50
Spacecraft/mission: STS-57
Crew: Ronald Grabe, Brian Duffy, G. David Low, Nancy Sherlock, Peter Wisoff, Janice Voss
Spacewalkers: G. David Low, Peter Wisoff
Purpose: EVA practice ahead of SSF assembly and HST repair; contingency EVA to secure Eureca antennas

Endeavour was scheduled at launch to remain in orbit for only 7 days. A 4 hr, 20 min EVA, the lowest priority major task on STS-57, was to be conducted only if the mission was extended to 8 days. Before the EVA, Endeavour's RMS snared ESA's European Retrievable Carrier (EURECA) satellite, which was launched on an Ariane rocket in 1992. The retrievable platform's antenna would not lock down, so flight controllers increased planned EVA duration to 5 hr, 50 min, and made securing the antenna the EVA's primary task. Low and Wisoff had to prebreathe for 4 hr because Spacehab module experiments required that Endeavour's cabin pressure be kept at 101.4 kpascal (14.7 psi) throughout the flight. Brian Duffy acted as IV crewman and Nancy Sherlock operated the RMS. Before the EVA the astronauts practiced translation in Endeavour's cramped middeck using the launch escape pole. They later judged that a single translation down the payload bay sill at EVA start was more effective practice. For the first time astronauts left the airlock through an extension which linked it to the Spacehab module in the payload bay. Low pushed on EURECA's antenna from the RMS while controllers on the ground commanded the latches to close. The antenna locked down, and the astronauts turned to the EVA DTOs. These were DTO 1210 and DTO 671 (EVA Hardware for Future Scheduled EVA Missions), which tested Hubble Space Telescope Servicing Mission-01 (HST SM-01) and Space Station EVA equipment. They took turns carrying each other while riding the RMS to judge their ability to move large loads, used a foot restraint while working with tools, and tested safety tethers. While away from the payload bay, pointed at space, the astronauts got cold enough to shiver, and their hands became numb and painful. In their postflight debriefing, Low and Wisoff called the DTOs "time well spent."

STS-57 Space Shuttle Mission Report, NSTS-08285, August 1993, pp. 27-28; memo, Randall McDaniel to EVA Section personnel (STS-57 EVA debrief notes), August 13, 1993; "STS-57: EVA Lessons Learned," presentation materials, Richard Fullerton, July 9, 1992.

July 1	STS-57/Endeavour landing
July 1	Mir/Soyuz-TM 17 PE-14 launch
July 22	Mir/Soyuz-TM 16 PE-13 landing
September 12	STS-51/Discovery launch

September 16
1993 EVA 5
World EVA 111
U.S. EVA 60

Shuttle EVA 22
Duration: 7:05
Spacecraft/mission: STS-51
Crew: Frank Culbertson, William Readdy, James Newman, Daniel Bursch, Carl Walz
Spacewalkers: James Newman, Carl Walz
Purpose: Test HST repair tools and procedures

Discovery's pilot, William Readdy, was IV crewmember; he helped Walz and Newman begin preparations for their EVA 30 min early. Immediately after leaving the airlock, they made a practice translation down the payload bay door sills. At the back of the bay Walz investigated damage caused by a pyrotechnic fastener malfunction in the Advanced Communications Technology Satellite payload. Walz found that the explosion had torn a hole in Discovery's aft payload bay bulkhead, but caused no damage to the payload bay doors or other equipment. He decided not to handle the debris because it might cut his gloves. The astronauts then commenced DTO 1210 and DTO 671 tests, tools for which were stored in the Provisional Stowage Assembly (PSA) compartment in the floor of Discovery's payload bay, next to the contingency payload bay door closure tools. Walz and Newman conducted a glove-warming evaluation using the payload bay lights, then tested tethers for high- and low-torque work and a Portable Foot Restraint (PFR) designed for HST SM-01. They also compared ground training with actual work on orbit. The astronauts reported that their WETF experience was more difficult than the actual EVA. The HST-related tests assured planners that ground preparations for the flight were on a sound footing. The astronauts accomplished more than planned and remained ahead of schedule until closeout, when a PSA door refused to close, so the EVA required 45 min more than scheduled. In their debriefing the crew stressed the importance of thermal vacuum tests as part of EVA testing and training.

STS-51 Space Shuttle Mission Report, NSTS-08286, December 1993, pp. 30-33; "Eleven Day Missions with EVA," Roelof Schuiling, *Spaceflight*, December 1993, pp. 425-426.

September 16
1993 EVA 6
World EVA 112
Russian EVA 52
Space Station EVA 60
Duration: 4:18
Spacecraft/mission: Mir PE-14
Crew: Vasili Tsibliyev, Alexandr Serebrov
Spacewalkers: Vasili Tsibliyev, Alexandr Serebrov
Purpose: Prepare for Rapana truss assembly

The Soviet Union planned to follow Mir with Mir-2, a large station building on Mir/Salyut hardware and incorporating a large truss to support solar dynamic power generation systems and antennas. Mir-2 was to have been resupplied in part by the Buran space shuttle, as well as advanced automated cargo vehicles based on Progress-M. The collapse of the Soviet Union in 1991 scuttled these plans, but paved the way for merging Mir-2 and SSF into one International Space Station (ISS). This EVA is often identified as Mir-2-related. Tsibliyev and Serebrov moved equipment from Kvant 2 to Kvant using Strela. They installed a "grate," then attached a platform behind Sofora. They then moved the container holding the Rapana truss to the attachment site and linked it to Mir's electrical system.

MirNews 187, Chris Vandenberg, September 17, 1993; MirNews 188, Chris Vandenberg, September 20, 1993; "Mir Spacewalks," *Spaceflight*, November 1993, p. 371; "Mir Mission Report," Neville Kidger, *Spaceflight*, May, 1994, p. 154.

September 20
1993 EVA 7
World EVA 113
Russian EVA 53
Space Station EVA 61
Duration: 3:13
Spacecraft/mission: Mir PE-14
Crew: Vasili Tsibliyev, Alexandr Serebrov
Spacewalkers: Vasili Tsibliyev, Alexandr Serebrov
Purpose: Rapana truss assembly experiment

This was the thirtieth EVA using Orlan-DMA space suits. Tsibliyev and Serebrov returned to the Kvant module to assemble the Rapana truss, a 26-kg (57.2-lb) cylindrical framework with memory alloy joints akin to those in Sofora. They expanded when heated, causing Rapana to unfold from its container. Extension to a length of 5 m (16.4 ft) required just 3 min. The truss was scheduled for analysis after 10 mo on Mir's exterior. The cosmonauts installed space exposure samples on Rapana before closing out the EVA.

MirNews 188, Chris Vandenberg, September 20, 1993; "Mir Spacewalks," *Spaceflight*, November 1993, p. 371; "Mir Mission Report," Neville Kidger, *Spaceflight*, May, 1994, p. 154.

September 22   STS-51/Discovery landing

September 28
1993 EVA 8
World EVA 114
Russian EVA 54
Space Station EVA 62
Duration: 1:52
Spacecraft/mission: Mir PE-14
Crew: Vasili Tsibliyev, Alexandr Serebrov
Spacewalkers: Vasili Tsibliyev, Alexandr Serebrov
Purpose: Inspect Mir's exterior (Panorama survey)

This EVA was the first connected with new U.S.-Russian cooperation in space. Originally scheduled for September 24, it was put back four days on September 22, possibly because of a SALC fault - on the day the delay was announced, the cosmonauts were reported to be repairing a pressure valve in Kvant 2. The primary objective of the EVA was the Panorama survey, a detailed inspection of Mir's exterior designed to provide Russian, U.S., and European engineers with assurance that Mir remained in good shape to support ambitious joint space projects after 7 yr in space. Panorama was also designed to assess damage caused by the intense Perseid meteor storm of August 1993. The EVA was planned to last more than 4 hr, but had to be cut short when Tsibliyev's suit cooling system failed. He stayed near the Kvant 2 airlock while Serebrov collected TREK experiment detector plates and videotaped and photographed Mir's exterior from Kvant 2. Serebrov, the first cosmonaut to test the SPK maneuvering device (1990), said later that the SPK should have been used for Panorama. On October 8, the Russians announced that an EVA to complete Panorama and perform other tasks would occur on October 22. They also announced that PE-14 would be extended because of problems in obtaining Soyuz booster engines to launch a replacement crew.

MirNews 189, Chris Vandenberg, September 22, 1993; MirNews 190, Chris Vandeberg, September 28, 1993; "Mir Mission Report," Neville Kidger, *Spaceflight*, May, 1994, p. 154.

October 22
1993 EVA 9
World EVA 115
Russian EVA 55
Space Station EVA 63
Duration: 0:38
Spacecraft/mission: Mir PE-14
Crew: Vasili Tsibliyev, Alexandr Serebrov
Spacewalkers: Vasili Tsibliyev, Alexandr Serebrov
Purpose: Inspect Mir's exterior (Panorama survey); miscellaneous tasks

With this EVA, scheduled to last 5 hr, Serebrov matched the old record for most career EVAs (9) held jointly by Leonid Kizim and Vladimir Solovyov since 1986. However, the EVA had to be terminated almost as it started because of a problem in the oxygen flow system of his Orlan-DMA suit, which had been worn 13 times by different cosmonauts and had exceeded its recommended operational lifetime. Before going inside the cosmonauts inspected and collected space exposure experiments, installed meteoroid detectors, and spoke with Russian Prime Minister Viktor Chernomyrdin, who was visiting the TsUP. They also snapped a few photos for the Panorama survey, but a third EVA had to be scheduled to complete the task.

MirNews 195, Chris Vandenberg, October 23, 1993; "Mir Mission Report," Neville Kidger, *Spaceflight*, May, 1994, p. 154; "The Experience in Operation and Improving the Orlan-type Space Suits," I. P. Abramov, *Acta Astronautica*, Vol. 36, No. 1, July 1995, pp. 1-12.

October 29
1993 EVA 10
World EVA 116
Russian EVA 56
Space Station EVA 64
Duration: 4:12
Spacecraft/mission: Mir PE-14
Crew: Vasili Tsibliyev, Alexandr Serebrov
Spacewalkers: Vasili Tsibliyev, Alexandr Serebrov
Purpose: Inspect Mir's exterior (Panorama survey); miscellaneous tasks

After two EVAs cut short by space suit problems, Tsibliyev and Serebrov at last completed the Panorama survey of Mir's exterior. They filmed Mir's main Altair/Luch geostationary satellite system communication antenna and the solar arrays. Then they inspected Sofora's mount and attached another space exposure cassette to Mir's exterior. The cosmonauts watched a piece of metal of undetermined origin drift past. Serebrov set a new career record for most EVAs (10), but Sergei Krikalev's record for most time spent in EVA (36 hr, 29 min) remained intact. The cosmonauts tossed the Orlan-DMA suit that gave trouble on the last EVA out the Kvant 2 hatch after rigging it so it appeared to be saluting. Analysis of the Panorama photos and videotape showed Mir's exterior to be intact but contaminated by thruster emissions. The solar arrays showed minor damage from meteoroids and orbital debris collisions.

MirNews 196, Chris Vandenberg, October 30, 1993; "Mir Mission Report," Neville Kidger, *Spaceflight*, May, 1994, p. 154; *Mir Hardware Heritage*, NASA RP 1357, David S. F. Portree, March 1995, p. 142.

December 2   **STS-61/Endeavour launch**

December 4
1993 EVA 11
World EVA 117
U.S. EVA 61
Shuttle EVA 23
Duration: 7:54
Spacecraft/mission: STS-61
Crew: Richard Covey, Ken Bowersox, Story Musgrave, Thomas Akers, Jeffrey Hoffman, Kathy Thornton, Claude Nicollier (ESA)
Spacewalkers: Story Musgrave, Jeffrey Hoffman
Purpose: Repair Hubble Space Telescope; prepare worksite, change gyroscopes, fuse plugs, and electronics control unit.

Endeavour's pre-dawn launch in pursuit of HST was visible for hundreds of kilometers up the east coast of the U.S. According to Story Musgrave, HST SM-01 was fraught with challenges; in addition to the night launch and "night shift" for the crew, STS-61 was planned as a grueling 11-day flight with five EVAs, the most for a Shuttle flight to date. Because of these challenges, the crew for the first Hubble servicing mission was made up of veterans. All four EVA astronauts had EVA experience; Richard Covey was a veteran Shuttle commander; Ken Bowersox was a veteran pilot; and European Space Agency (ESA) astronaut Claude Nicollier had operated the RMS in space before. According to Jeff Hoffman, this experience was particularly important when it came to preflight training - the crew all knew how training on the ground differed from actual space operations. The repair effort was subject to the attention of up to a dozen external review boards, one of which called for a backup EVA crewman. Before Gregory Harbaugh's assignment as backup, the EVA astronauts planned to fill in for each other if one became unavailable to fly. In the event, Harbaugh's services were not required, but his training experience made him ideally suited to act as HST SM-01 EVA CapCom. The HST servicing EVAs were developed over the course of 10 yr, but the astronauts made modifications based on their experience. For example, according to Hoffman, the original plan was to keep all tools outside, but the crew realized that "the critical consumable was EVA time." The astronauts decided that the hour spent preparing tools at the start of the EVA could be better spent starting the repair, so they kept some tools inside Endeavour's crew compartment so they could load up before they started using their EMU consumables. On December 3 Nicollier checked out the RMS while the EVA astronauts inspected their space suits, and cabin pressure was reduced from 101.4 kpascal (14.7 psi) to 70.3 kpascal (10.2 psi) to reduce prebreathe time. On this date, Covey guided Endeavour close to HST. Nicollier captured the telescope with the RMS and berthed it on the Flight Support System in Endeavour's aft payload bay. The primary EVA tasks included:

- replace solar arrays

- replace Rate Sensing Unit (RSU) 2

- replace Wide Field/Planetary Camera (WFPC) with improved WFPC-2

- replace Goddard High Speed Photometer (GHSP) with Corrective Optics Space Telescope Axial Replacement (COSTAR)

- replace Magnetic Sensing System (MSS) magnetometer 1

- replace RSU 3 and Electronics Control Unit (ECU) 3

- replace Solar Array Drive Electronics (SADE) 1 assembly

STS-61, 1993 - Jeffrey Hoffman and Story Musgrave (on RMS) service HST in Endeavour's payload bay. (STS061-98-050)

Secondary tasks included:

- install Goddard High-Resolution Spectrometer (GHRS) power supply redundancy kit

- install 386 co-processor in DF-224 primary computer

- replace MSS 2

- replace RSU fuse plugs

- replace ECU 1

Musgrave and Hoffman left Endeavour's airlock 1 hr early. They first installed a cover on a low-gain antenna to avoid damaging it, then changed RSU 2, RSU 3, ECU 1, ECU 3, and four fuse plugs. Replacements were stored in the Small Orbital Replacement Unit Protective Enclosure in the payload bay. Other replacement parts and tools for the EVAs were stowed in the ORU Carrier in the center of the payload bay. Endeavour carried 6545 kg (14,400 lb) of servicing gear, including more than 200 individual tools and crew aids. About 40 were for nominal repair operations, while the others were for contingencies. Musgrave, the shorter of the two astronauts, was able to slip under the HST sunshade to reach the RSUs. Because the astronauts did not have to remove the sunshade, this saved about 1 hr of EVA time. Musgrave unbolted the old RSU, then Hoffman, in a foot restraint on the RMS, slid it out while Musgrave steered it free. Up to this point, Hoffman and Musgrave were ahead of timeline, but then they had difficulty closing the RSU compartment door, which was deformed due to exposure to temperature extremes during HST's first 3 yr in space. Musgrave pushed on the bottom of the door while Hoffman, on the RMS, pushed on the top. This was the only major problem experienced on any of the HST SM-01 EVAs, a fact that surprised Hoffman, given the history of EVA. According to Hoffman, if the crew had not saved time by loading tools on the middeck and slipping under the sunshades, the EVA would have ended without the door closed and been deemed a failure. The astronauts closed out by setting up the payload bay for the next EVA.

"Hubble Repair Mission: STS-61," *Spaceflight*, January 1994, pp. 15-16; "STS-61 Mission Report," Roelof Schuiling, *Spaceflight*, March 1994, p. 78; *STS-61 Space Shuttle Mission Report*, NSTS-08288, February 1994, pp. 3, 5-6; interview, David S. F. Portree with Jeffrey Hoffman, June 18, 1996.

December 5

1993 EVA 12
World EVA 118
U.S. EVA 62
Shuttle EVA 24
Duration: 6:36
Spacecraft/mission: STS-61
Crew: Richard Covey, Ken Bowersox, Story Musgrave, Thomas Akers, Jeffrey Hoffman, Kathy Thornton, Claude Nicollier (ESA)
Spacewalkers: Kathy Thornton, Thomas Akers
Purpose: HST repair; replace solar arrays

Each of HST's British Aerospace-built solar array wings was 12 m (39.4 ft) long and 2.8 m (9.2 ft) wide fully deployed, weighed 160 kg (352 lb), and generated 2.5kW of electricity. Array replacement was scheduled for HST SM-01 before the telescope was launched. Originally this was meant to compensate for solar array degradation so total electricity produced would remain above 4.5kW, but problems with "jitters" in the arrays caused by thermal expansion and contraction quickly overshadowed the original justification. The new arrays were modified to eliminate the vibration problem. The arrays were stowed rolled up in a special Solar Array Carrier (SAC) at the front of the payload bay. After the first EVA the arrays were commanded to close. The starboard array failed to close properly because of a bent bistem in its support framework. The array was closed only 30 percent to avoid bistem breakage, which might have created a sharp-edge hazard. Thornton was unable to receive radio from Endeavour or the ground until 3 hr, 15 min into the EVA; she lost radio again near EVA's end. Akers served as relay during the blackout. The astronauts detached and jettisoned the starboard array, then installed its jitter-proof replacement. The port array was then removed and stowed in the SAC for return to Earth and its replacement installed. The new arrays were left rolled up. The astronauts finished the EVA by installing a foot restraint for the next EVA.

"Hubble Repair Mission: STS-61," *Spaceflight*, January 1994, p. 16; "STS-61 Mission Report," Roelof Schuiling, *Spaceflight*, March 1994, p. 78; "Hubble Space Telescope Servicing Mission Report No. 5, Hubble gets new ESA-supplied solar arrays," December 6, 1993; *STS-61 Space Shuttle Mission Report*, NSTS-08288, February 1994, pp. 3, 6; interview, David S. F. Portree with Kathy Thornton, June 17, 1996.

December 6
1993 EVA 13
World EVA 119
U.S. EVA 63
Shuttle EVA 25
Duration: 6:47
Spacecraft/mission: STS-61
Crew: Richard Covey, Ken Bowersox, Story Musgrave, Thomas Akers, Jeffrey Hoffman, Kathy Thornton, Claude Nicollier (ESA)
Spacewalkers: Story Musgrave, Jeffrey Hoffman
Purpose: Repair HST repair; replace WFPC with WFPC-2

The objective of the third HST servicing EVA was to replace the 281-kg (620-lb) Wide Field/ Planetary Camera. WFPC-2, built from a WFPC spare after ground controllers determined that HST had faulty optics, was stowed in the Radial Scientific Instrument Protective Enclosure (RSIPE). Musgrave and Hoffman removed the original WFPC and placed it in the RSIPE. The IV crew then tipped HST forward on the Flight Support System to put the closed aperture door within RMS reach. Both Hoffman and Musgrave mounted the RMS and replaced two of four MSS magnetometers near HST's aperture. They discovered that the MSS covers are disintegrating, raising fears that foam fragments from the covers might infiltrate HST's optics. The team installed four more fuse plugs and performed some light tasks originally scheduled for the next EVA before closing out.

"Hubble Repair Mission: STS-61," *Spaceflight*, January 1994, p. 16; "STS-61 Mission Report," Roelof Schuiling, *Spaceflight*, March 1994, p. 78; *STS-61 Space Shuttle Mission Report*, NSTS-08288, February 1994, pp. 3, 6; interview, David S. F. Portree with Jeffrey Hoffman, June 18, 1996.

December 7
1993 EVA 14
World EVA 120
U.S. EVA 64
Shuttle EVA 26
Duration: 6:50
Spacecraft/mission: STS-61
Crew: Richard Covey, Ken Bowersox, Story Musgrave, Thomas Akers, Jeffrey Hoffman, Kathy Thornton, Claude Nicollier (ESA)
Spacewalkers: Kathy Thornton, Thomas Akers
Purpose: Repair HST; install COSTAR unit

The GHSP had to be sacrificed to provide a place for the $50-million COSTAR device for correcting HST's faulty optics. Thornton mounted the RMS and opened the GHSP access door using a power rachet tool. One of the two JSC-supplied power tools carried on Endeavour failed completely, while the other lost its variable speed setting. Akers then climbed inside, disconnected the 221-kg (487-lb) photometer, pulled it out, and handed it to Thornton. Nicollier then moved Thornton and GHSP to a holding device, where she temporarily stowed the instrument. Thornton pulled the 290-kg (640-lb) COSTAR from the Axial Scientific Instrument Protective Enclosure

(ASIPE). Akers aligned the device on its rails, slid it in, engaged its latches, and connected its electrical cables. The astronauts then placed the GHSP in COSTAR's place in the ASIPE. The replacement task, scheduled to require 3 hr, 10 min, was completed in only 35 min. COSTAR could not be fully tested for 6 to 8 wk, but initial checks completed ahead of the final HST SM-01 EVA gave the device good marks. The astronauts then installed a new Goddard/Lockheed/ Fairchild DF-224 co-processor to improve computer performance. Communications problems continued for Thornton. Following the EVA, Covey and Bowersox fired Endeavour's forward thrusters for 61 sec to boost HST into a 516-km (321-mi) circular orbit.

STS-61 Space Shuttle Mission Report, NSTS-08288, February 1994, pp. 3, 5-6; "Hubble Repair Mission: STS-61," *Spaceflight*, January 1994, p. 16; "STS-61 Mission Report," Roelof Schuiling, *Spaceflight*, March 1994, p. 81; "Hubble Vision Problems Clear Up After December Repair," William Harwood, *Space News*, January 17-23, 1994; *Corrective Optics Space Telescope Axial Replacement*, Ball Corporation, October 1993; interview, David S. F. Portree with Kathy Thornton, June 17, 1996.

December 8
1993 EVA 15
World EVA 121
U.S. EVA 65
Shuttle EVA 27
Duration: 7:21
Spacecraft/mission: STS-61
Crew: Richard Covey, Kenneth Bowersox, Story Musgrave, Thomas Akers, Jeffrey Hoffman, Kathy Thornton, Claude Nicollier (ESA)
Spacewalkers: Story Musgrave, Jeffrey Hoffman
Purpose: Repair HST; replace solar array drive electronics

*STS-61, 1993 - Kathy Thornton (top) maneuvers the COSTAR unit toward Tom Akers.
(STS061-98-00Q)*

The astronauts replaced HST's SADE then put protective covers fabricated aboard Endeavour on MSS units 3 and 4 to prevent foam from getting into the telescope's optics. Controllers on the ground then commanded HST's solar arrays to unroll. They experienced trouble, so Hoffman and Musgrave aided deployment. Each array required 5 min to unroll. Communication problems meant no biomedical data during most of the EVA. The astronauts returned to the airlock, ending a record EVA total of 35 hr, 28 min for STS-61. According to Hoffman, by this EVA Nicollier could anticipate his needs. He would reach toward a tool or bolt and Nicollier would automatically move the RMS to put it within reach. Nicollier earned the nickname "the Magician" for his RMS acumen. On December 9 Nicollier used the RMS to grapple the telescope and lift it above Endeavour's payload bay. Ground controllers charged HST's batteries, deployed the high-gain antenna booms, and opened the aperture door. Nicollier then released the telescope. Covey and Bowersox moved Endeavour away, taking care not to strike HST with plumes from the orbiter's steering jets. On December 11 the crew performed space suit evaluations in the middeck and stowed the RMS. Endeavour landed in Florida on December 13.

STS-61 Space Shuttle Mission Report, NSTS-08288, February 1994, pp. 3, 5; "STS-61 Mission Report," Roelof Schuiling, *Spaceflight*, March 1994, p. 78; interview, David S. F. Portree with Jeffrey Hoffman, June 18, 1996.

December 13	**STS-61/Endeavour landing**

1994

January 8-July 9	**Mir/Soyuz-TM 18 PE-15**
January 14	**Mir/Soyuz-TM 17 PE-14 landing**
February 3-11	**STS-60/Discovery**
March 4-18	**STS-62/Columbia**
April 9-20	**STS-59/Endeavour**
July 1	**Mir/Soyuz-TM 19 PE-16 launch**
July 8-23	**STS-65/Columbia**
September 9	**STS-64/Discovery launch**

September 9
1994 EVA 1
World EVA 122
Russian EVA 57
Space Station EVA 65
Duration: 5:04
Spacecraft/mission: Mir PE-16
Crew: Yuri Malenchenko, Talgat Musabayev, Valeri Polyakov
Spacewalkers: Yuri Malenchenko, Talgat Musabayev
Purpose: Inspect docking port struck by Progress-M 24; mend thermal blanket torn by Soyuz-TM 17

Soyuz-TM 17 and Progress-M 24 struck Mir in January 1993 and August 1994, respectively. Valeri Polyakov, a medical doctor, monitored Malenchenko and Musabayev from inside Mir during this inspection EVA, which was scheduled to last 3 hr, 40 min. The cosmonauts started by replacing space exposure cassettes outside Kvant 2, then moved to the front of the Mir base block. They inspected Kristall near where it joined the base block and found that the damage caused by Soyuz-TM 17 was light - a missing 30-cm-by-40-cm (11.8-in-by-15.8-in) thermal blanket and scratches on insulation nearby. They mended the damage, then inspected the Progress-M 24 impact area on the Mir transfer compartment, finding no appreciable damage.

MirNews 229, Chris Vandenberg, September 13, 1994; "Mir Undamaged by Progress Bump," *Aviation Week & Space Technology*, September 19, 1994, p. 71; "Space Station Cosmonauts," Neville Kidger, *Spaceflight*, January 1995, pp. 8-9.

September 14
1994 EVA 2
World EVA 123
Russian EVA 58
Space Station EVA 66
Duration: 6:01
Spacecraft/mission: Mir PE-16
Crew: Yuri Malenchenko, Talgat Musabayev, Valeri Polyakov
Spacewalkers: Yuri Malenchenko, Talgat Musabayev
Purpose: Routine maintenance; prepare to move solar arrays from Kristall to Kvant; inspect Sofora

PE-16's second EVA, planned to last 4 hr, launched Mir's reconfiguration ahead of the scheduled arrival of the U.S. Shuttle Atlantis in June 1995. Malenchenko and Musabayev inspected the movable solar arrays on Kristall, which were scheduled to be retracted and moved to Kvant through a series of EVAs. They also inspected mounting brackets and drives on Kvant on which the arrays would be mounted. The cosmonauts took down space exposure cassettes from Rapana and inspected Sofora, then mounted a new amateur radio antenna. Valeri Polyakov tested the new antenna from inside Mir.

MirNews 229, September 13, 1994; "Mir Undamaged by Progress Bump," *Aviation Week & Space Technology*, September 19, 1994, p. 71; "Space Station Cosmonauts," Neville Kidger, *Spaceflight*, January 1995, p. 9.

September 16
1994 EVA 3
World EVA 124
U.S. EVA 66
Shuttle EVA 28
Duration: 6:51
Spacecraft/mission: STS-64
Crew: Richard Richards, L. Blaine Hammond, Carl Meade, Mark Lee, Susan Helms, Jerry Linenger
Spacewalkers: Mark Lee, Carl Meade
Purpose: Test SAFER self-rescue device

Mark Lee and Carl Meade performed the first untethered EVA since the MMU flights in 1984 to test the Simplified Aid for EVA Rescue (SAFER) device. Lee and Meade participated in SAFER air-bearing floor tests, virtual reality training development, and reach-and-access design tests, as

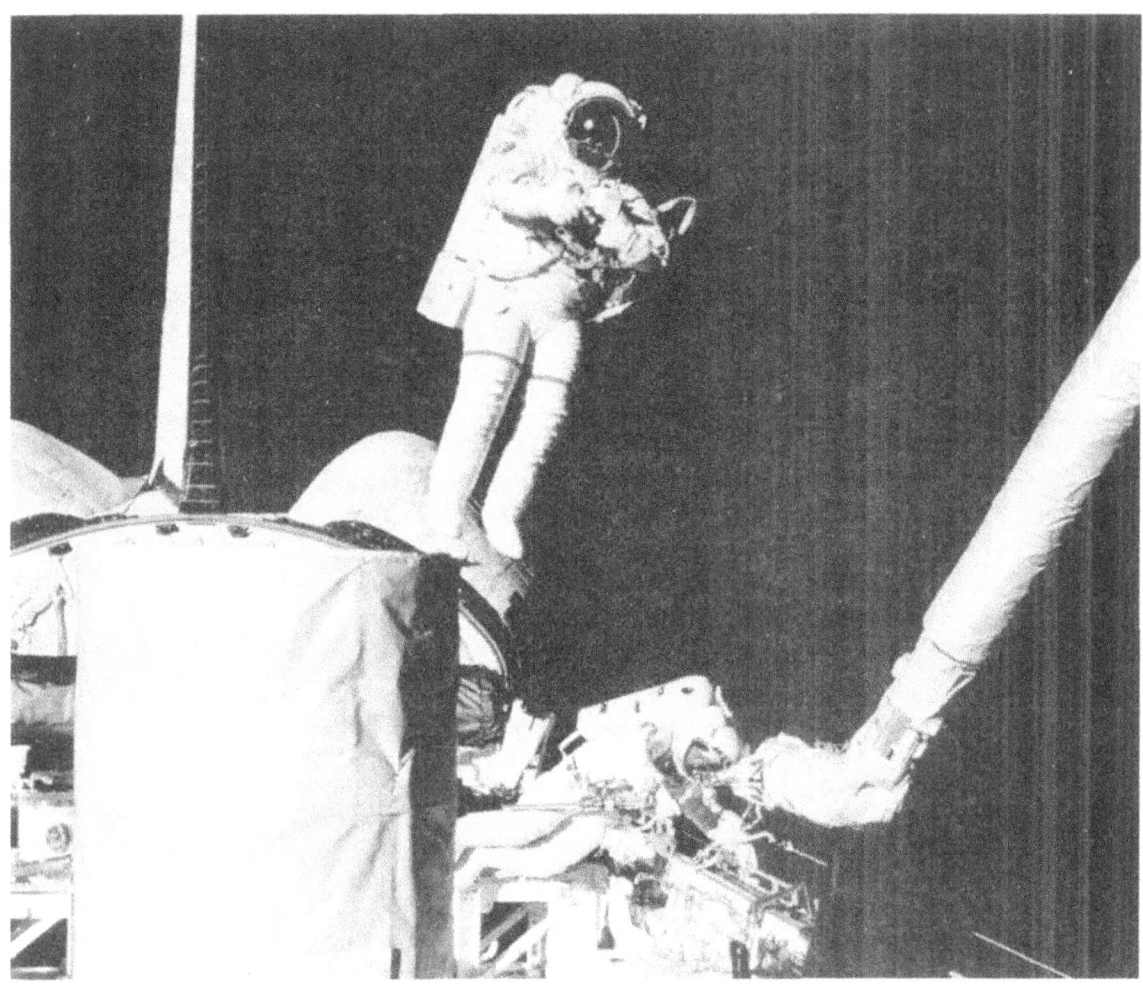

STS-64, 1994 - During the second EVA Development Flight Test mission, Mark Lee (top) and Carl Meade tested the SAFER outside Discovery's payload bay. (S064-115-011)

well as briefings to Shuttle Program management to secure development funding when funding from the Space Station program dried up for a time in 1993. SAFER is worn under the astronaut's PLSS backpack. The device weighs 37.7 kg (83 lb), more than 114 kg (250 lb) lighter than the MMU. SAFER attaches to the astronaut using the six existing PLSS hard points, including the two provided for the MMU, so no EMU modifications were required. "Towers" for thrusters extend up the sides of the PLSS. SAFER has no bulky hand controller arms in front. The hand controller is a small box - for this first test it was hard-secured to the astronaut's chest. In the ISS production model SAFER, the hand controller will be embedded in one of the thruster towers. The astronaut will pull a lanyard so the controller swings out on an arm, placing it within easy reach. SAFER has 24 fixed-position thrusters. Four compressed nitrogen tanks hold 60 sec of nitrogen if the device is used for translation, or 120 sec if it is used for rotation and stabilization. Jerry Linenger was IV crewman, while Susan Helms piloted the RMS. Commander Richard Richards and Pilot Blaine Hammond stood by to rescue the SAFER astronaut. The test had four parts - familiarization, a programmed jet test series to gather engineering data, tumbling tests, and precision maneuvering - and occurred within a 7.6-m-by-8.2-m (25-ft-by-27-ft) box of space at the front of Discovery's payload bay. The astronauts quickly determined that the device used less nitrogen than predicted. After completing the familiarization and engineering data portions of the test, Lee and Meade took turns standing in an MFR on the RMS and tumbling the other. The tumbled astronaut activated SAFER's automatic attitude hold system, stabilized, then maneuvered toward the RMS, which Helms pulled away to simulate a separation rate of 0.06 mps

(0.2 fps). Meade rolled Lee at 2 rpm - faster than planned, but SAFER stabilized him without difficulty. Finally, the astronauts took turns flying SAFER precisely along the RMS to a point near the aft flight deck windows. During the tests the astronauts replenished SAFER's propellant supply from the nitrogen recharge unit at the front of Discovery's payload bay seven times. The Electronic Cuff Checklist (ECC), a planned replacement for paper checklists, performed poorly, and Meade's feet became very cold.

STS-64 Space Shuttle Mission Report, NSTS-08293, January 1995, pp. 35-36; "Rescue Device Shines in Untethered EVA Tests," James McKenna, *Aviation Week & Space Technology*, September 26, 1994, pp. 25-26; interview, David S. F. Portree with Clifford Hess, May 30, 1996; interview, David S. F. Portree with Mark Lee, June 10, 1996.

September 20	**STS-64/Discovery landing**
September 30-October 11	**STS-68/Endeavour**
October 3	**Mir/Soyuz-TM 20 PE-17 launch**
November 3-14	**STS-66/Atlantis**
November 4	**Mir/Soyuz-TM 19 PE-16 landing**

1995

February 3	**STS-63/Discovery launch**

February 9
1995 EVA 1
World EVA 125
U.S. EVA 67
Shuttle EVA 29
Duration: 4:39
Spacecraft/mission: STS-63
Crew: James Wetherbee, Eileen Collins, Michael Foale, Janice Voss, Bernard Harris, Vladimir Titov (Russian Space Agency)
Spacewalkers: Michael Foale, Bernard Harris
Purpose: Mass handling and Shuttle EMU thermal evaluations in preparation for ISS assembly and maintenance

This was the first EVA in the EVA Development Flight Test (EDFT) program, which aimed to prepare NASA for ISS assembly. Foale and Harris commenced EVA preparations on February 8, soon after Discovery completed proximity operations with Russia's Mir space station. The astronauts wore EMUs with thermal modifications, including thicker underwear, better-insulated gloves, and a bypass switch to allow reduction of cooling water flow through the LCVGs without reducing ventilation. A 4-hr prebreathe was required because the crew could not lower cabin pressure prior to the EVA without compromising experiments in the Spacehab module in Discovery's payload bay. Foale and Harris waited in the airlock while Janice Voss retrieved the 1.37-m (4.5-ft) cube-shaped SPARTAN 204 freeflyer with the RMS. A few minutes before Voss berthed the 1363-kg (3000-lb) subsatellite, Foale and Harris entered the payload bay through the airlock extension tunnel linking Discovery's mid-deck to Spacehab. The astronauts secured two Universal Handling Tools on the SPARTAN for mass-handling experiments, then Harris and

Foale stepped onto the RMS. Russian cosmonaut Vladimir Titov took charge of the RMS and moved the astronauts 9.1 m (30 ft) above the payload bay for a 15-min cold soak. Discovery's attitude was purposely maintained to make the astronauts as cold as possible - the payload bay was pointed away from the Sun during daylight and toward deep space during orbital night. A "thermal cube" sensor package on the MFR and sensors in the EMU gloves recorded ambient temperatures so they could be compared with the astronauts' subjective impressions. The astronauts were not cold when the payload bay pointed toward the Earth. Then Foale moved to the PFR and Harris to a foot restraint by SPARTAN's port side. Voss unlatched the SPARTAN and Titov moved Foale into position over the freeflyer so he could lift it in an experiment to determine astronaut and EMU ability to handle large loads. Foale then handed the SPARTAN to Harris. (He commented later that the air-bearing floor simulated EVA mass-handling well.) During orbital night Foale's glove temperature dropped below minus 6 deg C (20 deg F). The ECCs went blank from the cold, so the astronauts fell back on printed checklists. Harris' feet became cold through contact with Discovery's structure as a temperature of minus 148 deg C (minus 130 deg F) was recorded by the thermal cube. The astronauts reported that the thermal overgloves produced only slight warming. Foale took back the SPARTAN, but Mission Control canceled the remainder of the mass-handling experiment and terminated the EVA early after the astronauts rated the cold as a 3 ("unacceptably cold") on a 1 to 8 scale devised before launch. Foale put the SPARTAN back in its berth. Commander James Wetherbee and Pilot Eileen Collins maneuvered Discovery to warm Harris and Foale's EMUs before they returned to the airlock. After taking off his helmet, Harris smelled an odor and suffered burning eyes. Wetherbee collected an air sample and Harris washed his eyes with water. Postflight analysis revealed no contaminants. The irritation could have been caused by contact with an anti-fogging soap solution - four EVA astronauts encountered the same problem on previous flights. During the EVA, a pressure drop occurred in the Spacehab module, which was isolated from the rest of Discovery's pressurized volume. The leak, equivalent to a loss of 7.7-9 kg (17-20 lb) of air per day, stopped after the EVA.

STS-63 Mission Report, June 1995; "EVAs Follow Mir Rendezvous," Spaceflight, May 1995, pp. 155-156; "STS-63 Mission Overview," Joel Powell, Countdown, March/April 1995, pp. 18-21.

February 11	**STS-63/Discovery landing**
March 2-18	**STS-67/Endeavour**
March 14	**Soyuz-TM 21/Mir PE-18 launch**
March 22	**Soyuz-TM 20/Mir PE-17 landing**

May 12
1995 EVA 2
World EVA 126
Russian EVA 59
Space Station EVA 67
Duration: 6:08
Spacecraft/mission: Mir PE-18
Crew: Vladimir Dezhurov, Gennadi Strekalov, Norman Thagard (NASA)
Spacewalkers: Vladimir Dezhurov, Gennadi Strekalov
Purpose: Prepare to move solar arrays from Kristall to Kvant

For a time it appeared that the first EVA of Mir PE-18 might be delayed by injury; Strekalov cut his hand while working inside the Mir station, and the cut became infected. Rumor had it that

U.S. astronaut Thagard might replace him, but this was vehemently denied by TsUP spokesmen, who stated that Thagard lacked specialized EVA training. On May 2 the cosmonauts inventoried and marked cables for use in the EVA. On May 5 they reviewed a training video, and on May 6 and May 7 prepared their Orlan-DMA space suits. On May 7 and May 8 the cosmonauts and TsUP performed communications checks. Dezhurov and Strekalov conducted a tool familiarization simulation in the depressurized Mir transfer compartment on May 10, during which a problem with the radio transmitter in one suit surfaced. The cosmonauts spent the four days prior to the EVA almost exclusively on EVA preparations. Thagard assisted and did life sciences research. The joint U.S.-Russian crew took breaks to load Progress-M 27 with trash prior to its scheduled undocking on May 21. On this date all was at last ready for the EVA, which was planned to last 5 hr, 20 min. Thagard, inside Mir, read instructions to Dezhurov and Strekalov when the station was out of contact with the TsUP. The cosmonauts changed wiring on Kvant to prepare for transfer of the 12.2-m (40-ft) Kristall arrays, then moved to Kristall and practiced folding three small panels of one array. Each array had 28 such panels. Removal of the U.S.-built TREK space exposure experiment (a 20-min procedure) was planned, but had to be postponed when the EVA ran 15 min past the safe limit. The cosmonauts were reported to be very tired after the EVA, so they rested all day May 13.

MirNews 252, Chris Vandenberg, May 8, 1995; MirNews 254, Chris Vandenberg, May 16, 1995; "Orbital Castling: Life of Mir Station Will Be Prolonged Three More Years," Mikhail Chernyshov, *Segodnya*, May 11, 1995, p. 9 (translated in *JPRS Report, Science & Technology, Central Eurasia: Space*, FBIS-UST-95-030, August 2, 1995, pp. 20-21); "Mir Extension Weekly Operations Report for Week of Monday May 1 through Friday, May 5, 1995," Anthony Sang to Distribution, May 9, 1995; "Mir-18 Mission Status Report #18," May 5, 1995; "Mir-18 Mission Status Report #20," May 12, 1995; "Mir Docking Plans Meet With Success," Neville Kidger, *Spaceflight*, July 1995, pp. 222-223.

May 17
1995 EVA 3
World EVA 127
Russian EVA 60
Space Station EVA 68
Duration: 6:52
Spacecraft/mission: Mir PE-18
Crew: Vladimir Dezhurov, Gennadi Strekalov, Norman Thagard (NASA)
Spacewalkers: Vladimir Dezhurov, Gennadi Strekalov
Purpose: Install solar array from Kristall on Kvant

On May 15 Strekalov and Dezhurov changed their suit batteries and replenished consumables. On this date Thagard cycled closure servomotor switches inside Mir while Strekalov monitored array closure from Strela, which Dezhurov operated. Strekalov had to manually close one panel. Detaching the array from Kristall required no tools; it was designed for easy removal by cosmonauts encumbered by space suit gloves. Dezhurov transferred Strekalov and the array to the Kvant worksite, then joined him there. Batteries for the arrays remained in Kristall, so cables had to run Mir's length. The cosmonauts lacked sufficient time to install the array and reopen it as planned. Instead they lashed it to its mount with tool tethers. The resulting power deficit was made up in part by electricity from the arrays on Progress-M 27, which was retained for two days longer than originally planned. The next scheduled EVA was delayed two days from May 20 to give the cosmonauts more time to rest.

MirNews 255, Chris Vandenberg, May 17, 1995; "EVA Problems Delay Mir Station Changes," Craig Covault, *Aviation Week & Space Technology*, May 22, 1995, p. 65; "Re: MEAT Weekly Reports 5 and 6," Bonnie Dunbar to Anthony Sang, April 27, 1995; "Mir-18: Public Affairs Office Mission Science Report," May 15, 1995; "Mir Docking Plans Meet With Success," Neville Kidger, *Spaceflight*, July 1995, p. 224.

May 22
1995 EVA 4
World EVA 128
Russian EVA 61
Space Station EVA 69
Duration: 5:15
Spacecraft/mission: Mir PE-18
Crew: Vladimir Dezhurov, Gennadi Strekalov, Norman Thagard (NASA)
Spacewalkers: Vladimir Dezhurov, Gennadi Strekalov
Purpose: Install Kristall array on Kvant; stow second Kristall array

The third EVA of Principal Expedition 18 (designated EVA "2B" by NASA) was scheduled to last 6 hr, 15 min. An electricity shortage interfered with communication between Mir and the TsUP during the EVA. There was insufficient power for TV through the Altair geosynchronous satellite. Despite this, Strekalov and Dezhurov succeeded in installing the array moved on May 17. Thagard then commanded the array to unfold, restoring Mir's electrical supply. The spacewalkers closed 13 of 28 segments on the second array so it could continue to produce electricity while leaving sufficient clearance for Kristall to be repositioned. Progress-M 27 undocked from the front port on May 23 to make way for Spektr.

MirNews 257, Chris Vandenberg, May 22, 1995; "Re: MEAT Weekly Reports 5 and 6," Bonnie Dunbar to Anthony Sang, April 27, 1995; "Mir Docking Plans Meet With Success," Neville Kidger, *Spaceflight*, July 1995, pp. 224; "A Mir Matter of Change," L. van den Abeelen, *Spaceflight*, August 1995, p. 273.

May 29
1995 EVA 5
World EVA 129
Russian EVA 62
Space Station EVA 70
Duration: 0:21
Spacecraft/mission: Mir PE-18
Crew: Vladimir Dezhurov, Gennadi Strekalov, Norman Thagard (NASA)
Spacewalkers: Vladimir Dezhurov, Gennadi Strekalov
Purpose: Prepare Mir base block berthing node for transfer of Kristall to -Z port

The cosmonauts wore space suits but remained inside the forward transfer compartment and never ventured outside Mir. Nevertheless, most sources consider this and the other Konus transfers to be EVAs. Dezhurov and Strekalov entered the transfer compartment, sealed the hatches linking it to Kristall, the core module, and Kvant 2, and dumped its air. They then removed the Konus #2 drogue from the -Y port, closed the port with a hinged flat plate door, opened an identical door on the -Z port, and installed the Konus unit. Their task completed, they repressurized the transfer compartment and rejoined Thagard inside the base block.

MirNews 257, Chris Vandenberg, May 22, 1995; "Mir Docking Plans Meet With Success," Neville Kidger, *Spaceflight*, July 1995, pp. 224.

June 2
1995 EVA 6
World EVA 130
Russian EVA 63
Space Station EVA 71

Duration: 0:23
Spacecraft/mission: Mir PE-18
Crew: Vladimir Dezhurov, Gennadi Strekalov, Norman Thagard (NASA)
Spacewalkers: Vladimir Dezhurov, Gennadi Strekalov
Purpose: Prepare Mir base block berthing node for transfer of Spektr module to berthed position

Dezhurov and Strekalov closed hatches leading to the Mir base block, Kvant 2, Kristall, and Spektr (which docked at the Mir front port on June 1), depressurized the transfer compartment, and moved Konus #2 to the -Y port so Spektr could be relocated there. (The new module was pivoted into place on June 3.) On June 5 two Spektr arrays were deployed. One failed to open fully, producing 20 percent less electricity than expected. The Russians announced plans to mount an unrehearsed EVA on June 15 to fully open the array ahead of Atlantis' docking with Mir. The EVA was rescheduled to June 16, then canceled because Strekalov refused to take part. He contended that the EVA was unnecessary and made hazardous by inadequate preparation. Rookie Commander Dezhurov argued with Strekalov, but the veteran flight engineer was adamant. After they returned to Earth, Dezhurov and Strekalov were each fined the equivalent of $9000 - 15 percent of the fee they had contracted to receive for the mission. The amount of electrical power available was judged sufficient for Atlantis' STS-71 docking with Mir, however.

MirNews 263, Chris Vandenberg, June 15, 1995; MirNews 264, Chris Vandenberg, June 17, 1995; question-and-answer session with Baumann Moscow State Technical University graduate cosmonauts, Baumann Institute, Moscow, Russia, April 11, 1996; "Star Brothers Fined $9000," *Kosomolskaya Pravda*, October 28, 1995, p. 1 (translated in *JPRS Report, Science & Technology, Central Eurasia: Space*, FBIS-UST-95-048, November 30, 1995, p. 50); "Historic Mir Docking," Neville Kidger, *Spaceflight*, August 1995, pp. 274.

| June 27-July 7 | STS-71/Atlantis (Shuttle-Mir 1) |
| July 13-22 | STS-70/Discovery |

July 14
1995 EVA 7
World EVA 131
Russian EVA 64
Space Station EVA 72
Duration: 5:34
Spacecraft/mission: Mir PE-19
Crew: Anatoli Solovyov, Nikolai Budarin
Spacewalkers: Anatoli Solovyov, Nikolai Budarin
Purpose: Inspect -Z port; manually deploy malfunctioning Spektr solar array

Solovyov and Budarin arrived aboard Mir on STS-71, spelling Dezhurov, Strekalov, and Thagard, who returned to Earth on the Shuttle. Planned duration of this first PE-19 EVA was 5 hr, 15 min. The cosmonauts inspected the -Z port where Kristall was to be relocated for leaks and found nothing out of the ordinary. The transfer compartment had experienced an unexplained slow pressure loss a month before. Solovyov and Budarin then used Strela to reach the balky Spektr array, which they succeeded in opening using a NASA-built tool. Small lateral array sections remained oriented 90 deg from planned final position, but the electricity lost was judged insignificant and the Spektr array repair declared complete. The cosmonauts then moved to Kvant 2, where they inspected an antenna and a malfunctioning solar array drive motor.

"Three EVAs for Mir Duo," Neville Kidger, *Spaceflight*, September 1995, p. 311.

July 19
1995 EVA 8
World EVA 132
Russian EVA 65
Space Station EVA 73
Duration: 3:08
Spacecraft/mission: Mir PE-19
Crew: Anatoli Solovyov, Nikolai Budarin
Spacewalkers: Anatoli Solovyov, Nikolai Budarin
Purpose: Install MIRAS infrared spectrometer; retrieve TREK detector

The planned 5-hr, 38-min EVA to install the Mir Infrared Atmospheric Spectrometer (MIRAS) was cut short because the cooling system in Commander Solovyov's Orlan-DMA suit failed, forcing him to remain linked by an umbilical to the cooling system in the Kvant 2 airlock. Flight Engineer Budarin prepared equipment for MIRAS installation on the next EVA. He also retrieved the U.S.-built TREK detector from Kvant 2's surface. TREK, placed on Kvant 2 during a 1991 EVA, was originally scheduled for return to Earth in 1993, but this was postponed because of more pressing EVA demands. The cosmonauts had difficulty closing the airlock hatch. The 220-kg (484-lb), 2.5-m-long (8.2-ft-long) Belgian-French-Russian MIRAS was originally designed for a 1995 launch on the exterior of the Mir-2 space station core module. After the Mir-2 program was combined with NASA's station program in 1993, MIRAS was modified for launch inside the Spektr module and subsequent EVA installation on Spektr's exterior. Modifications included splitting the device into two parts and reducing its diameter so it could fit through Mir's passageways.

"Three EVAs for Mir Duo," *Spaceflight*, September 1995, p. 311; "EVAs, Station Changes Keep Cosmonauts Busy," James Asker, *Aviation Week & Space Technology*, July 24, 1995, pp. 62-63; MirNews 266, Chris Vandenberg, July 17, 1995; "A Mir Matter of Change," L. van den Abeelen, *Spaceflight*, August 1995, p. 273; "Space shuttle to pick up package for UC Berkeley Physicists," UC Berkeley press release, November 8, 1995.

July 21
1995 EVA 9
World EVA 133
Russian EVA 66
Space Station EVA 74
Duration: 5:35
Spacecraft/mission: Mir PE-19
Crew: Anatoli Solovyov, Nikolai Budarin
Spacewalkers: Anatoli Solovyov, Nikolai Budarin
Purpose: Install MIRAS infrared spectrometer

Solovyov and Budarin repaired Solovyov's suit in consultation with Zvezda engineers and immediately began preparations for another EVA to install MIRAS. The cosmonauts attached the MIRAS package to Strela, then Budarin maneuvered both Solovyov and MIRAS to the worksite on Spektr. The cosmonauts needed about 2 hr to install MIRAS using three clamps. French engineers claimed that assembling the two-part instrument was "the most complex operation ever carried out during EVA." The cosmonauts completed their work early, but controllers in the TsUP received no data from the instrument. While Solovyov and Budarin stood by outside Mir, controllers traced the problem to Spektr's data transmission system. The cosmonauts corrected the problem and MIRAS began returning data. Early in this EVA Solovyov broke Krikalev's 1992 record for total career EVA time. By the time he returned to the Kvant 2 SALC his total was 41 hr, 49 min.

"Three EVAs for Mir Duo," *Spaceflight*, September 1995, p. 311; MirNews 268, Chris Vandenberg, July 22, 1995; "MIRAS O.K. After EVA Suspense," Theo Pirard, *Spaceflight*, November 1995, p. 370; presentation by Anatoli Solovyov, JSC Open House, August 24, 1996.

September 3	**Soyuz-TM 22/Mir PE-20 launch**
September 7	**STS-69/Endeavour launch**
September 11	**Soyuz-TM 21/Mir PE-19 landing**

September 16
1995 EVA 10
World EVA 134
U.S. EVA 68
Shuttle EVA 30
Duration: 6:46
Spacecraft/mission: STS-69
Crew: David Walker, Kenneth Cockrell, James Voss, James Newman, Michael Gernhardt
Spacewalkers: James Voss, Mike Gernhardt
Purpose: Practice space station assembly and maintenance tasks; evaluated EMU thermal improvements

The second EDFT EVA was also the 30th EVA of the Space Shuttle program. Voss and Gernhardt readied their EMUs to support a Wake Shield Facility contingency EVA, but none was required. Suit checkout lasted 2 hr, 26 min, and prebreathe time was increased from 40 min to 50 min because cabin depressurization to 70.3 kpascal (10.2 psi) occurred less than 24 hr ahead of the EVA. Pilot Kenneth Cockrell served as IV crewman. Sensors on their boots and PLSS backpacks recorded temperatures. Astronauts installed thermal cube sensors on the RMS; other sensors recorded the temperature on a task board on the starboard side of the payload bay. The task board allowed the astronauts to test EVA tools on space station hardware. Voss and Gernhardt took turns removing ISS-type micrometeoroid/orbital debris shields and insulation blankets from the board; they also tested power tools on typical space station fasteners and manipulated ORU boxes, an antenna boom, electrical conduits, and tethers while restrained and free floating. Astronauts graded each task for difficulty. When not working at the task board, each astronaut was cold-soaked 9.1 m (30 ft) over the payload bay for 45 min while doing repetitive tool-handling tasks. Their gloves had fingertip heaters powered by 3.7v lithium batteries. Suit improvements meant that the astronauts stayed comfortably warm. Voss and Gernhardt also tested new EMU helmet lights, footholds, handholds, tethers, and the ECC.

STS-69 Space Shuttle Mission Report, December 1995; "Successful EVA, Landing Cap Troubled Shuttle Flight," *Aviation Week & Space Technology*, September 25, 1995, pp. 118-119; *Space Shuttle Mission STS-69 Press Kit*, August 1995, p. 42; "NASA's Two-Satellite Mission," Roelof Schuiling, *Spaceflight*, December 1995, p. 419.

| September 18 | **STS-69/Endeavour landing** |

October 20
1995 EVA 11
World EVA 135
ESA EVA 1/Russian EVA 67
Space Station EVA 75
Duration: 5:11
Spacecraft/mission: Mir PE-20
Crew: Yuri Gidzenko, Sergei Avdeyev, Thomas Reiter
Spacewalkers: Thomas Reiter, Sergei Avdeyev
Purpose: Perform first ESA EVA; install experiments on Spektr

On October 6, the Russian Space Agency (RSA), ESA, and RKK Energia agreed to extend the Euromir 95/PE-20 mission by 44 days (from 135 days to 179 days) because of Russian financial problems and a shortage of Soyuz launch vehicles. Extending the mission shifted Principal Expedition 21 launch processing costs to FY1996. The crew was advised of a possible extension before launch. Despite some reports, indications are that they welcomed the extension, not least because RSA offered ESA a second EVA for Reiter. On this first EVA by an ESA astronaut, Reiter led the way out the Kvant 2 hatch. In addition to being a guest-researcher, he was the first cosmonaut from outside the former Soviet Union to earn the title "Flight Engineer." This made the EVA the first by two flight engineers. Reiter climbed onto the end of the Strela boom, then Avdeyev handed him the payload bag, moved him to Spektr, and used Strela as a handrail to join him. The cosmonauts crawled to the opposite side of Spektr to reach the European Space Exposure Facility (ESEF)-1. Reiter threaded a tether from the payload bag through wire loops attached to pins on ESEF-1, then pulled the tether to release covers, exposing four attachment sites. The cosmonauts installed two dust collectors, a space environment monitoring package, and a control electronics box. The dust collectors had motorized covers operable from within Mir. One of the dust collectors remained open at all times unless a Shuttle, Soyuz-TM, or Progress-M was near the station, then it was closed to avoid spacecraft thruster contamination. The other was opened only when Earth passed through dust left behind by comets. Commander Yuri Gidzenko powered up the ESEF-1 instrument from inside Mir and verified that it was functioning as expected. Reiter and Avdeyev then moved to a second worksite 2 m (6.5 ft) away from ESEF-1, where they replaced exposure cartridges with cartridges delivered by Progress M-29.

Moscow Office Report #155, ANSER Center for International Aerospace Cooperation, October 20, 1995; ESA Press Release Number 42-95, October 13, 1995; "In Orbit They Still Do Not Know That Their Mission Has Been Extended: And ESA Prepares to Receive an Unexpected Gift From Their Russian Colleagues," Sergei Novikov, *Segodniya*, October 14, 1995, p. 1 (translated in *JPRS Report, Science & Technology, Central Eurasia: Space*, FBIS-UST-95-030, November 30, 1995, pp. 44-45); MirNews 276, Chris Vandenberg, October 19, 1995; "U.K. Scientists to Collect Cosmic Dust," RAS Press Release, October 20, 1995; "Possible Extension of EUROMIR Mission and ESA's First Spacewalk," ESA Press Release 42-95, October 13, 1995.

October 20-November 5   STS-73/Columbia

November 12-20   STS-74/Atlantis (Shuttle-Mir 2)

December 8
1995 EVA 12
World EVA 136
Russian EVA 68
Space Station EVA 76
Duration: 0:37

Spacecraft/mission: Mir PE-20
Crew: Yuri Gidzenko, Sergei Avdeyev, Thomas Reiter
Spacewalkers: Sergei Avdeyev, Yuri Gidzenko
Purpose: Transfer docking cone to receive Priroda module.

For the 50th EVA using the Orlan-DMA space suit, the cosmonauts entered the base block transfer compartment, sealed hatches leading into Soyuz-TM 22, Spektr, Kvant 2, Kristall, and the base block, and vented the atmosphere into space. They then transferred the Konus #2 docking drogue from the -Z port to the +Z port to receive the Priroda module. Reiter waited out the EVA in the Soyuz-TM 22 descent module.

"Mir Highlights: Crew Work On and Welcome in 1996," Neville Kidger, *Spaceflight*, March 1996, p. 93.

1996

January 11   **STS-72/Endeavour launch**

January 15
1996 EVA 1
World EVA 137
U.S. EVA 69
Shuttle EVA 31
Duration: 6:09
Spacecraft/mission: STS-72
Crew: Brian Duffy, Brent Jett, Leroy Chiao, Winston Scott, Koichi Wakata, Daniel Barry
Spacewalkers: Leroy Chiao, Daniel Barry
Purpose: Gain EVA experience for ISS assembly

The first EVA of 1996 commenced shortly after midnight Houston time. The planned six-and-a-half-hr EVA tested equipment for ISS assembly tasks. Chiao and Barry attached a PFR to the RMS, then unfolded and practiced attaching a rigid umbilical diagonally across Endeavour's payload bay. Rigid umbilicals will carry fluid and electrical lines between the modules and trusses in the U.S. segment of the ISS. The astronauts then assembled and tested the Lockheed Martin-built Portable Work Platform (PWP), which consisted of three components - the Temporary Equipment Restraint Aid, a holding place for ORUs; the Portable Foot Restraint Workstation Stanchion, which included two toolboards, sliding locks for holding tools, and two rigid tether sockets for ORU tethers; and the Articulating Portable Foot Restraint. They then installed the PWP on the RMS. Chiao and Barry both accidentally switched on their suit lights during the EVA. Barry told Chiao at the end of the EVA that "we'll come out together [again], Leroy. . . there's going to be plenty of work to do for Station." During the postflight debriefing, Chiao said that the ECC was a good idea poorly executed. Glare made it hard to read and it got in the way of suit controls. He summed up NASA's EVA emphasis during a TV interview by stating that, "we're testing the materials that are going to go into Station and we're testing techniques we will ultimately use to build that station. . . finally, we're training people to go out and do those techniques in the space environment."

"U.S., Japanese Crew Hone Orbital Repertoire," James McKenna, *Aviation Week & Space Technology*, January 22, 1996, p.29; "Station Tools Pass Endeavour Tests," William Harwood, *Space News*, January 22-28, 1996, p. 18; STS-72 EVA Crew Debrief, February 1, 1996; "Shuttle Crew to Flight-Test Lockheed Martin's Space Station Assembly Equipment," Lockheed Martin Public Relations Office, January 10, 1996; "Space Station Assembly Tests," Roelof Schuiling, *Spaceflight*, April 1996, pp. 118-119.

January 17
1996 EVA 2
World EVA 138
U.S. EVA 70
Shuttle EVA 32
Duration: 6:54
Spacecraft/mission: STS-72
Crew: Brian Duffy, Brent Jett, Leroy Chiao, Winston Scott, Koichi Wakata, Daniel Barry
Spacewalkers: Leroy Chiao, Winston Scott
Purpose: Gain EVA experience for ISS assembly; test EMU thermal modifications

Suit donning problems delayed this EVA by about one hr. Once outside, Scott mounted a foot restraint and held still for 35 min while Endeavour rolled to chill the payload bay. The temperature dropped to minus 122 deg C (minus 104 deg F), providing an extreme test of fingernail glove heaters and the coolant-loop bypass system. According to Scott, he was aware of the low temperature but remained comfortable. In the postflight debriefing he told EVA engineers that "if I felt that comfortable standing still, I think I'd have been even warmer as a result of being busy." The astronauts tested cable trays and clamps for use on ISS. Chiao stated in the postflight debrief that cold temps would make handling pressurized lines difficult. They reported that the clamps caused hand fatigue, and that large connectors were harder to use than small. The astronauts rocked back and forth on foot restraints to determine the level of stress likely to be placed on ISS structures during EVAs. They again accidentally switched on their suit lights; in daylight they were unable to see that they were on. *Space News* quoted lead EVA engineer Daryl Schuck as saying that all the data gathered in the practice EVAs "is going to be folded into the station design" to "make sure we've got a station that we're going to be able to build and maintain."

STS-72 EVA Crew Debrief, February 1, 1996; "Station Tools Pass Endeavour Tests," William Harwood, *Space News*, January 22-28, 1996, p. 18; "U.S., Japanese Crew Hone Orbital Repertoire," James McKenna, *Aviation Week & Space Technology*, January 22, 1996, p.29; "Space Station Assembly Tests," Roelof Schuiling, *Spaceflight*, April 1996, pp. 118-119.

January 20   **STS-72/Endeavour landing**

February 8
1996 EVA 3
World EVA 139
ESA EVA 2/Russian EVA 69
Space Station EVA 77
Duration: 3:06
Spacecraft/mission: Mir PE-20
Crew: Yuri Gidzenko, Sergei Avdeyev, Thomas Reiter
Spacewalkers: Thomas Reiter, Yuri Gidzenko
Purpose: Recover ESEF-1 dust collectors from Spektr's exterior; repair antenna on Kvant 2 solar array

Reiter and Gidzenko trained by radio for this EVA, which was expected to last 5 hr, 30 min. The cosmonauts first moved the SPK MMU outside Kvant 2 for permanent storage. The bulky device, which took up room in the Kvant 2 EVA airlock compartment, was a hindrance to EVAs since its two flights in 1990. Then Gidzenko crawled down Strela and took his place at its control cranks. He moved Reiter from Kvant 2 to Spektr, then crawled down Strela and around Spektr to join Reiter at the ESEF-1 worksite, where the German flight-engineer removed two 2-kg (4.4-lb) dust collectors. The dust collectors were originally to have been recovered during an all-Russian EVA during PE-21, but postponement of PE-20's return by 44 days allowed Reiter to retrieve the collectors he installed in October. The cosmonauts fell behind schedule but caught up by working during night passes. They were unable to remove a malfunctioning antenna on one of the solar arrays with the tools at their disposal, so the TsUP ordered them to cut short the EVA. The "Post-EVA TGIF Message" they sent by amateur radio to a computer bulletin board showed that their exuberance was undamaged by the setback:

> Everything went fine, we have done our work and retrieved two astrophysical cassettes. . . now we are getting more and more into preparations for our return. It's Friday, and we just would not feel all right if we would not remind you about the TGIF drinks. . . 19 days [until landing] and counting. . .

The recovered ESEF-1 collectors returned to Earth with Reiter aboard Soyuz-TM 22 on February 29.

ESA Press Release 08-96, February 2, 1996; Ben Huset, February 13, 1996; "Spacewalkers Outside Mir Retrieve Experiments," Peter de Selding, *Space News*, February 12-18, 1996, p. 3; *Moscow Office Report* #155, ANSER Center for International Aerospace Cooperation, October 20, 1995; "Science Tasks and EVA for Crew," Neville Kidger, *Spaceflight*, April 1996, p. 141.

February 21	Soyuz-TM 23/Mir PE-21 launch
February 22-March 9	STS-75/Columbia
February 29	Soyuz-TM 22/Mir PE-20 landing

March 15
1996 EVA 4
World EVA 140
Russian EVA 70
Space Station EVA 78
Duration: 5:51
Spacecraft/mission: Mir PE-21
Crew: Yuri Onufrienko, Yuri Usachev
Spacewalkers: Yuri Onufrienko, Yuri Usachev
Purpose: Install second Strela boom on Mir base block

The first Strela boom could only reach only Mir's -Z side, so a second Strela boom was installed to allow the cosmonauts to translate easily to the repositioned Kristall module. The astronauts attached the boom to brackets on the base block and extended it to its full 12-m (39.3-ft) length, then used it to return to the Kvant 2 SALC.

"America joins Russia in Permanent Human Presence in Space," Neville Kidger, *Spaceflight*, June 1996, p. 188.

| March 22 | STS-76/Atlantis (Shuttle-Mir 3) launch |

March 27
1996 EVA 5
World EVA 141
U.S. EVA 71

Shuttle EVA 33/Space Station EVA 79
Duration: 6:02
Spacecraft/mission: STS-76 (Shuttle-Mir 3)/Mir PE-21
Crew 1: Kevin Chilton, Richard Searfoss, Linda Godwin, Michael Clifford, Ronald Sega (STS-76)
Crew 2: Yuri Onufrienko, Yuri Usachev, Shannon Lucid (NASA) (PE-21)
Spacewalkers: Michael Clifford, Linda Godwin
Purpose: Install MEEP space exposure panels on Mir Docking Module; gain EVA experience for ISS assembly; test ISS common foot restraint and tether hooks

Clifford and Godwin performed the first U.S. EVA outside a space station in more than 22 yr. Their EVA was significant also for being the first conducted from a Shuttle orbiter docked to a space station. Atlantis and Mir together weighed 237,494 kg (522,487 lb), making the combination the heaviest human artifact ever assembled in space. According to Godwin, the WETF was not deep enough to hold a mockup of the Mir Docking Module (DM) atop the Shuttle docking unit, so the astronauts "had to pretend a lot in training." For example, when they did egress simulations, they left the airlock, then were carried by divers to the mockup DM resting on its side. The astronauts used virtual reality for DM translation practice and SAFER training. Clifford wore the flight test SAFER unit used on STS-64, while Godwin wore a refurbished ground test unit. Before the EVA could commence, cargo for Mir stored in Atlantis' airlock had to be moved to Mir. For egress the crew used Atlantis' internal airlock plus the tunnel adapter leading to the docking unit, so they had available twice the usual airlock room. Godwin said that she could see the abandoned SPK hanging near the Kvant 2 hatch. The DM (delivered by Atlantis on STS-74) blocked some flight deck aft window views, but IV crewman Ron Sega was able to use payload bay cameras to monitor the EVA astronauts. Standard Shuttle tether hooks were not large enough to fit over Mir's handrails, so Godwin and Clifford used new tether hooks large enough to fit over Russian handrails and a foot restraint designed to hold both Orlan and EMU boots. They relied on their improved EMU helmet lights at night because Mir provided little illumination other than flashing running lights. Atlantis' payload bay lights helped relieve some of the gloom. During the EVA they took care not to venture beyond the 4.6-m-long (15-ft-long) DM onto Kristall because the Russians feared that they might accidentally damage its delicate surface structures. According to Godwin, NASA would have felt the same way about Russians unfamiliar with the Shuttle working in the payload bay. According to Godwin, Mir's white thermal blankets were stained brown around its attitude control jets. The crew aboard Mir, which included U.S. cosmonaut-researcher Shannon Lucid (she transferred to the Mir PE-21 crew soon after Atlantis docked), shot photos and video of the EVA, which they gave to Atlantis' crew before departure. Godwin and Clifford first attached clamps on handrails on the outside of the DM, then attached four 0.6-sq-m (2-sq-ft) Mir Environmental Effects Payload (MEEP) space exposure experiments to the clamps. One of the panels was designed to record the frequency of orbital debris impacts on Mir, while a second captured debris particles for later analysis. The last two contained 1000 samples of paint, fibers, optical and metallic coatings, insulation, and other materials to help U.S. ISS designers choose materials. MEEP would be removed and returned to Earth during joint U.S.-Russian EVAs on STS-86, scheduled for September 1997. The astronauts also removed a camera from the DM for return to Earth. Mir and Atlantis were in a warm attitude because the station had to be oriented to keep its solar arrays pointed at the Sun. Godwin said that her hands became almost too warm - at one point she thought that she had turned on her glove heaters. "I was glad to see sunset," she said.

"Mir Spacewalk Set for Atlantis Crew," James Asker, *Aviation Week & Space Technology*, March 18, 1996, p. 61; "Shuttle-Mir Flight Sets Stage for Station Era," James McKenna, *Aviation Week & Space Technology*, April 1, 1996, p. 25; *The New York Times*, March 28, 1996, pp. A1, A13; "STS-76 Mission Control Center Status Report #10," March 27, 1996; "Third Mir Docking: STS-76," Roelof Schuiling, *Spaceflight*, July

1996, p. 227; interview, David S. F. Portree with Clifford Hess, May 30, 1996; interview, David S. F. Portree with Linda Godwin, June 13, 1996.

March 31   STS-76/Atlantis (Shuttle-Mir 3) landing

May 19-29   STS-77/Endeavour

May 21
1996 EVA 6
World EVA 142
Russian EVA 71
Space Station EVA 80
Duration: 5:20
Spacecraft/mission: Mir PE-21
Crew: Yuri Onufrienko, Yuri Usachev, Shannon Lucid (NASA)
Spacewalkers: Yuri Onufrienko, Yuri Usachev
Purpose: Transfer MCSA from Docking Module to Kvant installation site; begin installation

For this EVA, the first since her arrival on Mir, U.S. astronaut Shannon Lucid operated a video camera to record the cosmonauts' activities. Mir had relatively few viewports, preventing her from seeing the entire EVA. In a letter home to Earth, Lucid recounted the start of the EVA:

> I hear them exiting the airlock and leaving the station. . . no sooner were they out the airlock, than Yuri was yelling at me to look out the window and start taking pictures. I looked out and there was my commander perched on the end of a very long white pole [Strela] arcing over the blue and white Earth below. . . My first thought when I saw this was, "Wow, the future is now. This is real space station work." For a number of years now, I have been seeing artist renditions of what it would be like when the International Space Station is being worked on in a routine manner by astronauts, but this was no artistic fantasy; this was real life.

Commander Onufrienko and Flight Engineer Usachev used the Strela boom installed in March to move to the Docking Module at the end of Kristall. There they removed the Mir Cooperative Solar Array (MCSA), jointly developed by NASA and RSA, and delivered attached to the Docking Module by Atlantis on STS-74. NASA Lewis Research Center in Cleveland, Ohio managed the U.S. contribution to MCSA. The array was designed to help increase available Mir power, extending the station's life and supplying additional power for U.S. experiments. The cosmonauts secured MCSA to Strela, then moved it to Kvant, removed it from its container, and attached it to a mounting bracket. The cosmonauts returned to the Kvant 2 airlock using Strela and assembled a 1.2-m (3.9-ft) Pepsi can replica from aluminum struts and nylon sheets. The oversized replica was delivered by Progress M-31. They videotaped each other near the replica, then disassembled it for return to Earth. The videotape would be used by Pepsico in a commercial campaign.

"Lucid writes home about space walking protocol," *Space News Roundup*, NASA JSC, July 26, 1996, pp. 1, 4; "NASA Science comes to Mir," Neville Kidger, *Spaceflight*, September 1996, pp. 296; "Mir Cooperative Solar Array is Deployed/ISS Power Hardware Being Built and Tested," NASA press release 96-107, May 25, 1996.

May 24
1996 EVA 7
World EVA 143

Russian EVA 72
Space Station EVA 81
Duration: 5:34
Spacecraft/mission: Mir PE-21
Crew: Yuri Onufrienko, Yuri Usachev, Shannon Lucid (NASA)
Spacewalkers: Yuri Onufrienko, Yuri Usachev
Purpose: Install and deploy MCSA on Kvant

Onufrienko and Usachev deployed the 18-m-long (59-ft-long) MCSA outside the Kvant module using a handcrank. The array, which unfolded like an accordion, had 84 "panel modules" of 80 silicon solar cells each. The cells were identical to those planned for use on the U.S. segment of the International Space Station. The cosmonauts linked the array to Mir's power supply, but the electrical cables used permitted power to be supplied from only half of the array. Fully operational, MCSA would supply 6 kW of electricity - at the end of this EVA, it supplied half that. A Progress-M supply ship would deliver new cables for installation on a future spacewalk.

"Lucid Writes Home About Space Walking Protocol," *Space News Roundup*, NASA JSC, July 26, 1996, pp. 1, 4; "NASA Science Comes to Mir," Neville Kidger, *Spaceflight*, September 1996, pp. 296; "Mir Cooperative Solar Array is Deployed/ISS Power Hardware Being Built and Tested," NASA press release 96-107, May 25, 1996; "Life on Mir with the Cosmonauts," Neville Kidger, *Spaceflight*, April 1997, pp. 115-116.

May 30
1996 EVA 8
World EVA 144
Russian EVA 73
Space Station EVA 82
Duration: 4:20
Spacecraft/mission: Mir PE-21
Crew: Yuri Onufrienko, Yuri Usachev, Shannon Lucid (NASA)
Spacewalkers: Yuri Onufrienko, Yuri Usachev
Purpose: Install U.S.-built MOMS instrument outside Priroda module

The Modular Optoelectrical Multispectral Scanner (MOMS) device was launched inside Priroda for later EVA installation. MOMS flew on the STS-7 and STS 41-B Shuttle missions. Prior to the EVA, the cosmonauts planned their activities and positioned equipment in the airlock, then gathered together tools and placed them in their tool tray. To prepare for the EVA, which was scheduled for the middle of the night, the crew slept in, ate lunch, then slept again. Before entering the airlock, Commander Onufrienko placed a piece of red tape across controls Lucid was not to touch. While in the airlock he and Usachev asked her about station air pressure levels and the location of the complex over the Earth. Lucid found the EVA difficult to observe through Mir's few tiny ports. Radio communication with the EVA crew was excellent, Lucid reported, but there was little communication with the TsUP. The cosmonauts attached the MOMS to its site on Priroda's exterior. The cosmonauts referred to the male and female electrical connectors as "mamas" and "papas," which Lucid said made Mir feel "warm and homey."

"Lucid Writes Home About Space Walking Protocol," *Space News Roundup*, NASA JSC, July 26, 1996, pp. 1, 4; "NASA Science Comes to Mir," Neville Kidger, *Spaceflight*, September 1996, pp. 296.

June 6
1996 EVA 9
World EVA 145

Russian EVA 74
Space Station EVA 83
Duration: 3:34
Spacecraft/mission: Mir PE-21
Crew: Yuri Onufrienko, Yuri Usachev, Shannon Lucid (NASA)
Spacewalkers: Yuri Onufrienko, Yuri Usachev
Purpose: Install sample cassettes; install micrometeoroid detectors

The cosmonauts moved to Spektr and replaced cassettes in the Swiss/Russian Kozma experiment, then installed the Particle Impact Experiment (PIE) and the Mir Sample Return Experiment (MSRE). They then installed the SKK-11 cassette, which exposed construction materials to space conditions.

"Lucid Writes Home About Space Walking Protocol," *Space News Roundup*, NASA JSC, July 26, 1996, pp. 1, 4; "NASA Science Comes to Mir," Neville Kidger, *Spaceflight*, September 1996, pp. 296.

June 13
1996 EVA 10
World EVA 146
Russian EVA 75
Space Station EVA 84
Duration: 5:42
Spacecraft/mission: Mir PE-21
Crew: Yuri Onufrienko, Yuri Usachev, Shannon Lucid (NASA)
Spacewalkers: Yuri Onufrienko, Yuri Usachev
Purpose: Install Ferma-3 (Rapana) structure on Kvant; deploy Travers antenna

The Strombus (Ferma-3) was a 5.9-m (19.35-ft) truss made up of four sections. Onufrienko and Usachev assembled and installed the structure on Kvant's underside, then moved to Priroda and manually deployed the saddle-shaped Travers Synthetic Aperture Radar antenna. The cosmonauts closed out the last EVA of PE-21 by filming the final segment of the Pepsi commercial.

"Lucid Writes Home About Space Walking Protocol," *Space News Roundup*, NASA JSC, July 26, 1996, pp. 1, 4; "NASA Science Comes to Mir," Neville Kidger, *Spaceflight*, September 1996, pp. 296.

June 20-July 7	STS-78/Columbia
August 17	Soyuz-TM 24/Mir PE-22 launch
September 2	Soyuz-TM 23/Mir PE-21 landing
September 16-26	STS-79/Atlantis (Shuttle-Mir 4)
November 19-December 7	STS-80/Columbia

December 2
1996 EVA 11
World EVA 147
Russian EVA 76
Space Station EVA 85
Duration: 5:57
Spacecraft/mission: Mir PE-22

Crew: Valeri Korzun, Alexandr Kaleri, John Blaha (NASA)
Spacewalkers: Valeri Korzun, Alexandr Kaleri
Purpose: Install Mir Cooperative Solar Array cables; move Rapana truss

Korzun and Kaleri installed a 23-m (75.5-ft) cable to double to 6 kW the amount of electricity provided by MCSA. This involved attaching the cable to the array, then trailing it to the socket for the Mir base block "top" solar array, which was no longer used because it was shadowed by Kvant 2. The cosmonauts then moved the Rapana girder to the top of the new Strombus girder on Kvant's underside.

"Life on Mir with the Cosmonauts," Neville Kidger, *Spaceflight*, April 1997, pp. 115-116; MirNews 337, Chris Vandenberg, December 3, 1996.

December 9
1996 EVA 12
World EVA 148
Russian EVA 77
Space Station EVA 86
Duration: 6:36
Spacecraft/mission: Mir PE-22
Crew: Valeri Korzun, Alexandr Kaleri, John Blaha (NASA)
Spacewalkers: Valeri Korzun, Alexandr Kaleri
Purpose: Attach Kurs rendezvous radar antenna to Docking Module

Blaha, the third NASA astronaut to stay for an extended period aboard Mir, checked systems and executed Mir control commands provided him by the TsUP during this EVA, which lasted longer than planned. Korzun and Kaleri had difficulties handling cumbersome cable bundles when they installed a new omnidirectional Kurs antenna on the Docking Module. Before returning inside, they reattached a cable to an amateur radio antenna which they had knocked loose during their first EVA.

"Life on Mir with the Cosmonauts," Neville Kidger, *Spaceflight*, April 1997, p. 116; MirNews 338, Chris Vandenberg, December 9, 1996; "NASA 3/Mir 22 Status Report-13," December 13, 1996.

1997

January 12-22	STS-81/Atlantis (Shuttle-Mir 5)
February 10	Soyuz-TM 25/PE-23 launch
February 11	STS-82/Discovery launch

February 13
1997 EVA 1
World EVA 149
U.S. EVA 72
Shuttle EVA 34
Duration: 6:42
Spacecraft/mission: STS-82
Crew: Ken Bowersox, Scott Horowitz, Mark Lee, Greg Harbaugh, Steve Smith, Steve Hawley, Joe Tanner

Spacewalkers: Mark Lee, Steve Smith
Purpose: HST repair; replace FOS with NICMOS; replace GHRS with STIS

Steve Hawley grappled HST with the RMS early on this date, at 3:34 a.m. Houston time, and about 30 min later berthed it on the Flight Support System in Discovery's payload bay. He then used the RMS camera to survey the telescope's exterior, detecting an orbital debris crater on the aft shroud that "looks like a volcano." Hawley deployed HST after its launch aboard Discovery on STS-31 in April 1990. Late on this date, Mark Lee, an EVA veteran, and EVA rookie Steve Smith commenced the second series of HST servicing EVAs (HST SM-02). The astronauts had more than 150 tools. Two of the astronauts wore EMUs incorporating improvements for ISS use. Lee and Smith entered the airlock, which, during Discovery's just-completed maintenance period in California, was moved outside of the crew cabin. This was done to create more space in the middeck and to better position the airlock for ISS use. They left the airlock after a delay of several min caused when one of HST's twin 12.2-m (40-ft) solar arrays windmilled through a quarter turn and bounced back. Postflight analysis showed that air vented from the airlock was funneled through thermal blankets in Discovery's payload pay and vented onto the array. The crew reported that the event lasted only 5 sec. The first 17 deg of rotation required 14 sec, then the SADE automatically powered off, cutting telemetry. Total motion through 120 deg required 60 sec. In postflight debrief Lee accounted for the discrepancy between the inflight crew estimate and postflight analysis by saying that the crew witnessed only the last part of the event. Smith rode the RMS while Hawley operated it. The astronauts removed the GHRS and replaced it with the 318-kg (700-lb) Space Telescope Imaging Spectrograph (STIS). The GHRS had suffered partial electrical system failure on February 7. The astronauts noted that yellow paint was flaking off HST handrails, raising the possibility that they might contaminate its delicate inner workings. Lee and Smith then removed the Faint Object Spectrograph, the last of the original science instruments launched on HST, and replaced it with the Near Infrared Camera and Multi-Object Spectrometer (NICMOS). Ground controllers determined that the NICMOS and STIS instruments were operating and ready for their calibration period, which would last until May. (During the calibration period, controllers determined that NICMOS camera 3 could not be focused.) The EVA concluded at 6:17 a.m. Houston time on February 14.

"STS-82 - New Instruments for the HST," Roelof Schuiling, *Spaceflight*, June 1997, pp. 205-208; "Hubble Mission Scrambles to Make Surprise Repairs," Craig Covault, *Aviation Week & Space Technology*, February 24, 1997, pp. 20-23; STS-82 EVA crew debrief, March 5, 1997; "Shuttle Retrieves Hubble; Russians Launch to Mir," Craig Covault, *Aviation Week & Space Technology*, February 17, 1997, pp. 53-55; "Goddard Checks Hubble After Shuttle Servicing," Craig Covault, *Aviation Week & Space Technology*, March 3, 1997, pp. 24-25; *HST Project Crew Debrief Ques. List*, March 5, 1997.

February 14
1997 EVA 2
World EVA 150
U.S. EVA 73
Shuttle EVA 35
Duration: 7:27
Spacecraft/mission: STS-82
Crew: Ken Bowersox, Scott Horowitz, Mark Lee, Greg Harbaugh, Steve Smith, Steve Hawley, Joe Tanner
Spacewalkers: Greg Harbaugh, Joe Tanner
Purpose: HST repair; replace FGS, failed engineering and science tape recorder; install optical control electronics enhancement kit

The second EVA of the second HST servicing mission again included a veteran EVA astronaut

(Harbaugh) and a rookie (Tanner). Tanner rode the RMS. The astronauts replaced a worn-out 227-kg (500-lb) Fine Guidance Sensor with a modified spare and installed electronics enhancements that allowed it to operate properly, then replaced a failed data recorder with a solid-state recorder capable of simultaneous record and playback. The new recorder could hold ten times as much data as the original 1970s-vintage unit. During the EVA Commander Ken Bowersox (who was pilot for HST SM-02) and Pilot Scott Horowitz boosted the telescope's orbit 2.9 km (1.8 miles) using Discovery's small maneuvering thrusters. The astronauts detected an orbital debris impact scar on an HST antenna and scattered cracks in the teflon outer layer of HST's 17-layer thermal insulation blankets. They closed out the EVA at 5:52 a.m. Houston time on February 15.

"STS-82 - New Instruments for the HST," Roelof Schuiling, *Spaceflight*, June 1997, p. 208; "Jonathan's Space Report, No. 314," February 21, 1997; "Hubble Mission Scrambles to Make Surprise Repairs," Craig Covault, *Aviation Week & Space Technology*, February 24, 1997, pp. 20-23; STS-82 EVA crew debrief, March 5, 1997; "Shuttle Retrieves Hubble; Russians Launch to Mir," Craig Covault, *Aviation Week & Space Technology*, February 17, 1997, pp. 53-55; "Goddard Checks Hubble After Shuttle Servicing," Craig Covault, *Aviation Week & Space Technology*, March 3, 1997, pp. 24-25.

February 15
1997 EVA 3
World EVA 151
U.S. EVA 74
Shuttle EVA 36
Duration: 7:11
Spacecraft/mission: STS-82
Crew: Ken Bowersox, Scott Horowitz, Mark Lee, Greg Harbaugh, Steve Smith, Steve Hawley, Joe Tanner
Spacewalkers: Mark Lee, Steve Smith
Purpose: HST repair; replace data interface unit; enhance engineering and science tape recorder; replace reaction wheel assembly unit

Smith and Lee replaced an old reel-to-reel recorder with a new Solid State Recorder and replaced a worn-out Reaction Wheel Assembly unit, one of four used to point HST. Bowersox and Horowitz boosted HST to a slightly higher orbit during the EVA using maneuvering jets. Lee carefully examined the damaged insulation, stating that, "it's cracking all over the place." The EVA concluded at 5:04 a.m. on February 16. Mission managers then decided to add a fifth HST repair EVA on February 17 to attempt to mend damaged thermal insulation. The damage was concentrated on HST's sunward side.

"STS-82 - New Instruments for the HST," Roelof Schuiling, *Spaceflight*, June 1997, p. 208; "Hubble Mission Scrambles to Make Surprise Repairs," Craig Covault, *Aviation Week & Space Technology*, February 24, 1997, pp. 20-23; STS-82 EVA crew debrief, March 5, 1997; "Shuttle Retrieves Hubble; Russians Launch to Mir," Craig Covault, *Aviation Week & Space Technology*, February 17, 1997, pp. 53-55.

February 16
1997 EVA 4
World EVA 152
U.S. EVA 75
Shuttle EVA 36
Duration: 6:34
Spacecraft/mission: STS-82
Crew: Ken Bowersox, Scott Horowitz, Mark Lee, Greg Harbaugh, Steve Smith, Steve Hawley, Joe Tanner

Spacewalkers: Greg Harbaugh, Joe Tanner
Purpose: HST repair; replace SADE and magnetometer covers; install thermal blankets

In what was meant to be the final EVA of HST SM-02, astronauts replaced the jury-rigged magnetometer covers installed by Story Musgrave and Jeffrey Hoffman during the last EVA of HST SM-01, on December 8, 1993. They worked about 18 m (60 ft) "above" Discovery's payload bay. They also replaced a SADE unit, which involved handling bolts and connectors not designed for EVA replacement. They laced into place multi-layered insulation blankets over damaged insulation near HST's aperture. Damage included a long gash. The spare insulation was carried aboard Discovery for contingency repair of micrometeoroid damage and prepared in the middeck prior to the EVA. Instructions for preparing four patches for the next EVA were radioed to Lee and Smith inside Discovery during this EVA. Harbaugh and Tanner ended the EVA at 5:19 a.m. on February 17.

"STS-82 - New Instruments for the HST," Roelof Schuiling, *Spaceflight*, June 1997, p. 209; "Hubble Mission Scrambles to Make Surprise Repairs," Craig Covault, *Aviation Week & Space Technology*, February 24, 1997, pp. 20-23; STS-82 EVA crew debrief, March 5, 1997; "Shuttle Retrieves Hubble; Russians Launch to Mir," Craig Covault, *Aviation Week & Space Technology*, February 17, 1997, pp. 53-55.

February 17
1997 EVA 5
World EVA 153
U.S. EVA 76
Shuttle EVA 37
Duration: 5:17
Spacecraft/mission: STS-82
Crew: Ken Bowersox, Don Horowitz, Mark Lee, Greg Harbaugh, Steve Smith, Steve Hawley, Joe Tanner
Spacewalkers: Mark Lee, Steve Smith
Purpose: HST repair; install thermal blanket patches

Because they had a limited supply of insulation, the astronauts placed thermal blanket patches on only three parts of the telescope where insulation had begun to curl away from the metal structure, leaving other areas for future servicing missions. The astronauts rigged vertical wires to hold the blankets in place. Their work completed, they waited in the airlock while controllers investigated a possible problem with one of HST's Reaction Wheel Assemblies. Controllers briefly considered adding a sixth EVA to replace the component - a spare was carried on board - but they concluded that no repair was necessary. The EVA ended at 3:32 a.m. on February 18. Bowersox and Horowitz boosted HST's orbit by 4.8 km (3 mi), to a final release orbit of 536 by 514 km (335 by 321 mi). This was a record operating altitude for HST. Hawley released the telescope with the RMS on February 19. HST Project Scientist Edward Weiler told *Aviation Week & Space Technology* that "we don't have the original Hubble Space Telescope any more - we've got a new telescope. You can call it Hubble 2."

"STS-82 - New Instruments for the HST," Roelof Schuiling, *Spaceflight*, June 1997, p. 209; "Hubble Mission Scrambles to Make Surprise Repairs," Craig Covault, *Aviation Week & Space Technology*, February 24, 1997, pp. 20-23; STS-82 EVA crew debrief, March 5, 1997; "Shuttle Retrieves Hubble; Russians Launch to Mir," Craig Covault, *Aviation Week & Space Technology*, February 17, 1997, pp. 53-55.

February 21   **STS-82/Discovery landing**

March 2   **Soyuz-TM 24/Mir PE-22 landing**

April 29
1997 EVA 6
World EVA 154
Russian EVA 78/U.S. EVA 77
Space Station EVA 87
Duration: 4:57
Spacecraft/mission: Mir PE-23
Crew: Vasili Tsibliyev, Alexandr Lazutkin, Jerry Linenger (NASA)
Spacewalkers: Vasili Tsibliyev, Jerry Linenger
Purpose: Test Orlan-M space suit; deploy Optical Properties Monitor; retrieve PIE and MSRE; install radiation dosimeter

Jerry Linenger reached Mir aboard Atlantis on mission STS-81, spelling John Blaha. The fourth NASA astronaut to live on the station, Linenger contended with an oxygen system fire on February 23 and coolant system leaks which sprayed ethylene glycol into the station's atmosphere. Russian sources stated that the ethylene glycol concentration in Mir's air neared dangerous levels in mid-April. The station's troubles overshadowed this EVA, delaying it from April 17. For a time it appeared that the EVA might be canceled. On this date Tsibliyev and Linenger donned the new Orlan-M space suits delivered on Progress M-34. Orlan-M constituted a modest upgrade of the Orlan-DMA - the most noticeable addition was a second visor on the top of the helmet. Linenger and Tsibliyev used Strela to move to the Kristall module to install the Optical Properties Monitor, then returned to Kvant 2 to remove the U.S. MSRE and PIE, which returned to Earth with Linenger. They finished by installing the Benton radiation dosimeter. During the EVA, Lazutkin videotaped Tsibliyev and Linenger and monitored Mir's systems. In his postflight EVA debriefing Linenger stated that his crew received the go-ahead to do the EVA with mixed feelings - they were, after all, very busy trying to keep the station functional. He then described various elements of Russian EVA-related hardware:

Orlan-M - The new Orlan variant performed "beautifully" - its 40.7 kpascal (5.9 psi) operating pressure was not a hindrance, nor did it produce fatigue. Tsibliyev reported that the new Orlan-M gloves were easier to use than the Orlan-DMA gloves. Both Tsibliyev and Linenger chose to wear U.S. comfort gloves. During orbital night the suit became cold. Tsibliyev pulled a muscle while struggling out of the suit after the EVA. The suit "took an incredible amount of prep. . . it was like rebuilding an engine."

Kvant 2 EVA hatch - According to Linenger, this was "never adequately repaired" after Solovyov and Balandin damaged it in 1990. Instead, it was "jerry-rigged" with C-clamps. Tsibliyev was nervous about handling the hatch and forbade Linenger to touch it.

Strela - When Tsibliyev slewed the boom toward Kristall with Linenger on the end, it bent "like a fishing pole," bouncing Linenger back and forth "like a yo-yo." The boom was not effective for precision positioning.

Mir's exterior - The Mir station's hull is "a tangle of all kinds of junk," with unmarked dead-end traverse routes, delicate solar arrays popping out at all angles, discarded equipment, disused experiments and experiment mounts, and built-in sharp-edge hazards. None of these, said Linenger, appear on the training mockups in the Star City Hydrolaboratory. Linenger saw no sign of "road sign" markers reportedly installed on earlier EVAs.

NASA-Mir 4, 1997 - Wearing a new Orlan-M space suit, Jerry Linenger translates up the Strela boom outside Russia's Mir space station. Mir-23 commander Vladimir Tsibliyev is the photographer. Orlan-M consitutes an Orlan upgrade for Russian participation in the International Space Station program. Note the helmet's oval "top" visor and rear-entry hatch. (NM23-43-002)

According to Linenger, based on his experience, Russian EVA differs in many ways from U.S. EVA. These differences include:

- EVA training is general and not task-specific. According to Linenger, the philosophy of Russian EVA trainers appeared to be, "we've trained you, you know how to use your suit, just go do it."

- EVA timeline is of reduced importance - however, the cosmonauts would occasionally work through orbital night to make up for lost time. During night work they use their helmet visor lights, which are adequate for work at one spot but not for finding one's way over the hull.

- More free floating - fewer handholds, footholds, and tether points.

- Because Mir consists of generally convex surfaces (unlike the Shuttle orbiter with its concave payload bay), during EVA outside Mir "you feel like you're falling at 18,000 miles per hour."

- Little documentation - suit donning procedures are summed up in a large cardboard flipbook attached to the Kvant 2 airlock wall. The cosmonauts lack written timelines and task lists.

- No interaction between the IV crew and the EVA crew - Lazutkin did not communicate with Tsibliyev and Linenger during the EVA.

After returning inside Mir, Linenger described his feelings during the EVA in an email letter to his son:

> You are. . . on a cliff. Crawling, slithering, gripping, reaching. . . the whole cliff is falling and you are on it. . . it is difficult to discount the feeling that you are moving away, detached. In the midst of all this, you carry out your work calmly, methodically. You snap a picture or two, and below notice the Straits of Gibraltar. . .

Linenger returned to Earth aboard Atlantis on May 24, 1997, ending a 132-day stay in space.

"U.S. Astronaut Ready for Milestone Spacewalk," NASA press release 97-80, April 25, 1997; NASA 4 (Jerry Linenger) EVA debrief, June 6, 1997; "Shuttle-Mir Web: Letters to My Son," April 29-30, 1997 (http://shuttle-mir.nasa.gov/shuttle-mir/); "Russian, U.S. spacemen begin space walk," Reuter report, April 29, 1997; "Antifreeze Leak at Mir Station Has Reached Maximum Level," Reuters report, April 22, 1997; "Linenger Completes Space Walk," *Space News Roundup*, May 9, 1997, pp. 1, 8; "Astronaut and Cosmonaut Join in Space Excursion," *Space News*, May 5-11, 1997, p. 17.

Index

Monographs in Aerospace History

Launius, Roger D., and Gillette, Aaron K. Compilers. *The Space Shuttle: An Annotated Bibliography*. (Monographs in Aerospace History, No. 1, 1992).

Launius, Roger D., and Hunley, J.D. Compilers. *An Annotated Bibliography of the Apollo Program*. (Monographs in Aerospace History, No. 2, 1994).

Launius, Roger D. *Apollo: A Retrospective Analysis*. (Monographs in Aerospace History, No. 3, 1994).

Hansen, James R. *Enchanted Rendezvous: John C. Houbolt and the Genesis of the Lunar-Orbit Rendezvous Concept*. (Monographs in Aerospace History, No. 4, 1995).

Gorn, Michael H. *Hugh L. Dryden's Career in Aviation and Space*. (Monographs in Aerospace History, No. 5, 1996).

Powers, Sheryll Goecke. *Women in Aeronautical Engineering at the Dryden Flight Research Center, 1946-1994* (Monographs in Aerospace History, No. 6, 1997).

www.ingramcontent.com/pod-product-compliance
Lightning Source LLC
Chambersburg PA
CBHW081455170526
45166CB00008B/2433